Barns Across America

Heber Bouland

Published by the
American Society of Agricultural Engineers
2950 Niles Road
St. Joseph, Michigan

About ASAE — the Society for engineering in agricultural, food, and biological systems

ASAE is a technical and professional organization of members committed to improving agriculture through engineering. Many of our 8,000 members in the United Stated, Canada, and more than 100 other countries are engineering professionals actively involved in designing the farm equipment that continues to help the world's farmers feed the growing population. We're proud of the triumphs of the agriculture and equipment industry. ASAE is dedicated to preserving the record of this progress for others. This book joins many other popular ASAE titles in recording the exciting developments in agricultural history.

Barns Across America

Editor and Book Designer: Melissa Carpenter Miller

Library of Congress Catalog Card Number (LCCN) 98-73792
International Standard Book Number (ISBN) 0-929355-96-2

Printed in the U.S.A.

ABOUT THE AUTHOR

Heber Bouland is both an agricultural and civil engineer and has worked in both government and private practice. He has done research, has designed and evaluated a variety of agricultural structures, and has written many articles and papers related to farming and agriculture. He has managed agricultural relief and rural development projects in the Philippines, Central America, and Africa. He is a registered professional engineer in Maryland.

One of Mr. Bouland's hobbies is photography. All of the pictures shown in this book without a source were taken by him.

Mr. Bouland is interested in preservation of barns and other of our nation's historic treasures. He serves on the Montgomery County Maryland Rustic Roads Advisory Committee and is a member of the American Society of Agricultural Engineers and the National Trust for Historic Preservation. He currently resides in Ashton, Maryland and Loma Linda, California.

TO DOLORES

CONTENTS

ACKNOWLEDGEMENTS

This book would not have been possible without the help of many people. People at the various state and local historical societies and historic preservation offices helped me find great barns to visit and study. I want to particularly acknowledge: Patrick Zollner (formerly at Arkansas), William J. Macintire (Kentucky), Gerald Lee Gilleard (Missouri), Suzanne Fisher (Indiana), Clare Cavicchi (Montgomery County, Maryland), Susan G. Pearl (Prince George's County, Maryland) Cate Whitmore (Riverside County, California), and Carl Holm (Cumberland County, New Jersey).

I also received valuable help from university professors and agricultural extension agents. John Morgan at Emory and Henry College and Ray Bucklin at the University of Florida were especially helpful.

Librarians and other specialists at various libraries helped me find photographs, illustrations, and valuable barn information. These librarians were at the National Agricultural Library, the Library of Congress, and the National Archives. I want to give special thanks to Betty Branch of the Photographic Division, USDA; Dace Taube of the University Library, University of Southern California; and Jeff Wyatt, the National Registry of Historic Places, National Park Service.

Various family members — Robert, Kelli, and Janet — helped by either solving computer problems, typing, or reviewing the manuscript. I want to especially thank my wife, Dolores, for her invaluable assistance in doing research, typing, and commenting on drafts.

Some people helped formulate the scope of the book by reviewing various outlines and proposals: James Hedrick, agricultural economist; the late Harold Thompson, agricultural engineer; Donald Ortner, anthropologist; and Gordon Laing, retired book editor.

Many experts reviewed drafts of selected parts of the manuscript. A better product has emerged because of their review. They were: Don Blayney, USDA dairy expert; Philip Dole, Professor Emeritus, University of Oregon; Bob Ensminger, author; Mary Staley, Professor of fine arts, Montgomery College; Skip Miller, Historian, Kit Carson Museum, Taos, New Mexico; Allen Nobel, Professor of Geography, University of Akron; Terry Sharrer, Curator of Medicine and formerly Curator of Agriculture, National Museum of American History, Smithsonian Institution; and Lowell Soike, Iowa Historical Society.

I particularly want to credit the following who carefully reviewed the entire manuscript: Wayne Price, barn enthusiast, Springfield, Illinois; James Whitaker, agricultural engineer and author, Storrs, Connecticut; and Robert Yeck, Silver Spring, Maryland, former Chief, Farm Structures and Livestock Branch, USDA.

I want to give thanks to Donna Hull and Melissa Miller who edited and put the book together. Finally, I want to give special thanks to the farmers who let me visit and photograph their barns and who took their time to tell me barn stories.

I would like to acknowledge the contribution of many talented photographers:

Bruce Brittian, Ithaca, New York
Dennis Buffington, University Park, Pennsylvania
Eugene H. Lambert, Redlands, California
M.D. Locke Jr., Milwaukie, Oregon
David Orcutt, Redlands, California
John Prucell, Boyds, Maryland
John L.Telford, Silver Spring, Maryland
James T. Walker, Bowie, Maryland
Llene Wellman, Lodi, California
Fred Wilson, Massillon, Ohio

To all of the above I say thank you.

INTRODUCTION

This book tells the story of barns and the important role they have played in American life. It covers five centuries of American history and the 3,000 miles separating the oceans. The story begins 500 years ago when this land was first called America and Indians stored their maize and other food in clay jars, baskets, and crude cribs. Animal shelters did not exist, except for turkey and parrot cages; the Indians didn't keep domestic livestock.

When Europeans arrived, they brought cows, horses, burros, sheep, pigs, cereal grains, and fruits. Like Noah's ark, ships carried the beginnings of new animal and plant kingdoms to America. At first the European settlers didn't shelter their livestock. They built hovels to store food and protect only their most prized animals in the severest climates. Soon settlers brought their European building traditions to America, erecting barns with massive timber frames, hand-hewn with the axe, and with large floors for threshing grain. Settlers built many different types of barns often reflecting the builder's ethnic origin. Today we refer to these barns by such names as: English, Dutch, Swedish, German, poteaux-en-terre, and casa-corral. In the Mid-Atlantic colonies, many diverse groups settled, creating a strong agriculture and building sturdy barns.

During the late 19th century, as more sawed lumber became available, farmers erected lighter frame wood barns. Advancements in germ theory, haying, grain harvesting, and animal husbandry also changed the barn. Grain threshing became a field operation and barns were used mostly to shelter animals and store their forage. Farmers built bigger barns with better ventilation, more windows, and larger hay lofts.

For the most part, the movement of barn building was westward with simple temporary structures leading the way as the frontier advanced. Barns started as part of the village or fort and later shifted to the countryside setting. Moving across the American landscape from the Atlantic to the Pacific, one sees barns in a variety of fasci-nating sizes and materials. Local weather conditions, topography, and farming operations molded designs. Builders continued to take their ethnic barn building traditions west. Barns vary from the cozy New England connected barn, built of heavy timber, to Midwestern barns with lighter framing and low sweeping roofs to the open cool adobe sheds and simple corrals of the Southwest.

Today many barns are large food factories built of steel, composite wood products, and reinforced concrete. Thousands of livestock or poultry are raised in one building complex, complete with feed storage bins, environmental controls, and elaborate waste disposal systems. Yesterday's barns were designed and built by farmers or local craftsmen using materials found on the farm. Today, professional engineers and architects design farm buildings prefabricated in factories far from the farm, and operated by large commercial enterprises. Yesterday's barns had charm and quaint-ness; today's barns, efficiency.

Monuments to Our Rural Heritage, the first chapter, covers 500 years of barn building. It starts with simple Indian food storage and ends with high-tech barns for the 21st century. These monu-ments include an elegant round stone barn built by the Shakers in the early 1800s that tells us of their simple religious lifestyle; their handiwork was an expression of worship. Frontier log and sod barns are symbols of a powerful pioneering spirit.

After the barn history, I present a chapter called *Life Down at the Barn*. This chapter describes barn life primarily during the late 19th and early 20th centuries, when electricity and the internal combus-tion engine hadn't yet fully replaced horses and manual labor. Cows and horses shuffled about, interrupted by the morning and evening chores of milking and feeding the livestock. Periodically, however, neighbors or outsiders came with strange machinery to do a big farm job, dramatically interrupting the daily routine. Peddlers came to wheel and deal; preachers to hold fervent religious meetings.

Barns have been the scenes for the naughty and the romantic. From time to time, renowned events happened in a barn, the shooting of a major fugitive, the start of a great city fire, and for some traditions, the birthplace of a great religious leader.

Designs for the Times, the third chapter, gives more details on five great American barns showing how they met the needs of their time and region: log, German bank, adobe, round, and modern dairy. For example, log barns were ideal for Appalachia, the South, and other regions where timber was abundant and tools were limited mostly to the axe. In the Southwest, ranchers mixed earth and water to raise great haciendas out of the earth. Today, great cow cities in Arizona and New Mexico, having thousands of milk cows, provide milk to markets hundreds of miles away.

Barns have innate beauty because they are genuine. For the most part, they don't have gaudy colors or showy decorations. Many barns aren't even painted. It's the plain honesty of the barn that inspires. *Barnscapes*, the final chapter, covers barn beauty and aesthetics. Older barns have a quiet beauty as seen in sun-bleached decaying wood; they are like a fine piece of antique furniture. Artists appreciate the graceful roof lines, functional shapes, rhythmic patterns, simple decorations, and warm textures. Early barns were usually built of wood and stone, ideal subjects for pencil or ink drawings. This is why we often see barns on calendars, greeting cards, and note pads. We look at how several artists see barns and the farmstead so differently. There are the grim, somber paintings of Andrew Wyeth, the cheerful fun-filled paintings of Grandma Moses, and the storybook pictures of Grant Wood.

The book's goal is to help you discover the historical richness of barns, understand barn life, and appreciate the vernacular beauty of barns across America. While it's not a technical book, it should be of interest to technical people: agricultural engineers, architects, agricultural and architectural historians, folklorists, anthropologists, and structural engineers. Barn lovers everywhere should enjoy the book: farmers, retired farmers, anyone who grew up on the farm, those who visited Grandma and Grandpa's farm during the summer, people interested in visual arts, history buffs, and city folks with a yearning for country life.

Research for this book is based on scouring many books and periodicals published from the 1800s to today. I interviewed farmers and barn builders, explored many barns, and attended lectures and exhibits on barns such as The Barn Again exhibit held at the National Building Museum, Washington, DC, in 1994. I talked to agricultural and architectural historians from historical societies and government historic preservation offices.

I took many photographs in the book. Other photographs and drawings came from the Library of Congress, the National Archives, the National Gallery of Art, State Offices of Historic Preservation, historical societies, United States Department of Agriculture, and freelance photographers.

The important role the barn has played in American life is shown by the frequency the word is used in figures of speech. We use expressions such as: "Broad as the side of a barn," "Locking the barn door after the horse was stolen," and "Barn storming across the country." Mothers ask the critical question of their teenage children: "Were you born in a barn?" Retail merchants try to show they have a large selection and honest country trading policies by calling their store a barn. There are dress barns, boot barns, tire barns, and mattress barns. The images are both good and bad. In any event, these vanishing landmarks remind us of older customs and farming practices, and are symbols of how farmers lived — their courage, strength, and persistence.

I hope you enjoy the pictures and text, and that they help you find the legacy barns have left us and enhance your awareness of barn life and beauty.

— HB

Barns Across America

Heber Bouland

MONUMENTS TO OUR RURAL HERITAGE

This epic event has elements of the Noah's Ark story. When the early Spanish ships arrived in the Americas in the early 1500s they carried cows, burros, sheep, and seed grain — all of which would become the genetic pool for a new kingdom of domestic animals and plants. In addition, the Spanish brought European farming methods to what they thought was a new world. These new farming methods dramatically changed forever the growing and storage techniques. Eventually it led to the creation of traditional barns in America.

THATCHED CRIBS, ADOBE BINS: 1500-1607

When the Spanish started exploring North America in the early 1500s, it marked the beginning of the end of thousands of years of Indian dominance over the land and marked the beginning of traditional Western agriculture and barn building in America. Europeans used a crude plow pulled by draft animals. They raised sheep, swine, and cattle that were confined, sheltered, fed, and watered. The Spanish built thatched palm buildings for communal food storehouses in Florida and other parts of what is now the southeastern United States. By the end of the 16th century the Spanish had a permanent settlement in New Mexico where they built simple buildings of brush and mud, and later built grand adobe ranch buildings.

Growing and Storage Methods

When the Europeans arrived in the Americas there were at least one million Indians north of the Rio Grande; about three-fourths were living in what is now the continental United States. Hundreds of tribes with many different languages, lifestyles, crafts, shelters, and economies lived on this land. Indians relied on three methods for obtaining food: (1) primarily hunting, fishing, and gathering;

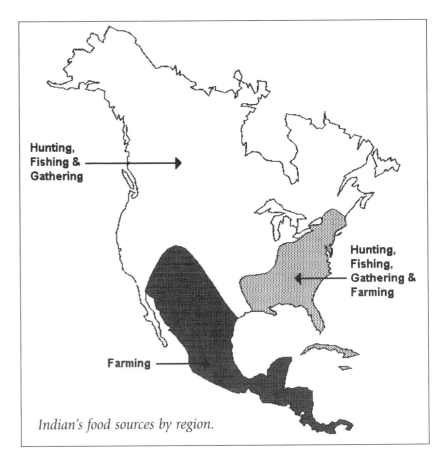

Indian's food sources by region.

(2) hunting, fishing, and gathering, plus farming; (3) primarily farming.

The Indians stored food in clay pots, baskets, animal skins, adobe bins, and small wood thatched cribs — the same methods their ancestors had used for thousands of years. Pottery and baskets filled

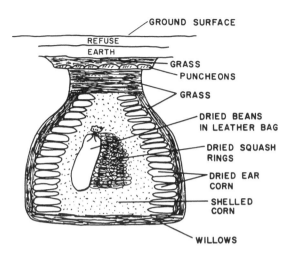

GROUND SURFACE
REFUSE
EARTH
GRASS
PUNCHEONS
GRASS
DRIED BEANS IN LEATHER BAG
DRIED SQUASH RINGS
DRIED EAR CORN
SHELLED CORN
WILLOWS

Many tribes stored dried corn, beans, and squash in bell-shaped pits about 3 feet deep and 4 feet in diameter. (Source: Steve Ahlar, after work of Frederick Wilson)

with corn, nuts, dried meat, or berries lay on the floor in the corner or hung from poles that framed the structure of the tepee or wigwam where the Indians lived. Indians sometimes stored food in underground pits to hide it when they moved. The thatched cribs and adobe bins, along with the pottery, baskets, and pits, could be considered the first American barns. A barn is a building for storing farm products or feed and for housing animals. These early methods don't strictly meet a barn's definition because they were sometimes used to store food gained from gathering and hunting — not from farming. The Indians didn't provide shelters for their animals; their only domesticated animals were dogs, turkeys, and parrots.

Regional Variations

Indians in the central Great Plains and the Pacific Northwest hunted, fished, or gathered their food. On the Plains they hunted buffalo. They built temporary shelters, such as the familiar tall cone-shaped tepees of wood poles covered with buffalo hides, that could easily be rolled up and moved. Indians in the Columbia River Basin and along the Pacific Coast depended on fishing. They built wood plank houses and hung fish from the rafters. The Indians in northern California gathered acorns and stored them in cribs raised off the ground and made of pole framing covered with leaves and bark.

In the Eastern Woodlands, primarily the land east of the Mississippi River, the Indians not only gathered nuts and berries, fished, and hunted small game, but also planted corn, pumpkins, squash, beans, and tobacco. They slashed and burned the woods and planted their crops around the remaining tree trunks. They erected dome-shaped wigwams which were more permanent structures than the hunters used. These wood-frame structures were covered with bark, reeds, or palm leaves. Some southeastern Indians built corn cribs — pole-framed buildings raised off the ground and covered with leaves, reeds, or palm leaves. Each family had its own crib and there were public granaries for storing food for the chiefs, visitors, and for emergencies.

The Indians that most heavily relied on farming for food were in the Southwest, specifically where the states of Arizona, Colorado, New Mexico, and Utah meet. The Pueblo Indians grew corn, cotton, beans, and squash. They planted in the river valleys and may have used irrigation and had the most permanent food storage and shelters. The Pueblos built structures as high as five stories and were

The Southeastern Muskhogeans Indians stored food in elevated cylindrical thatched cribs like the small one at the right of this village. (Source: Copyright © 1959 by Edwin Tunis. Revised copyright © by Maryland National Bank, Executor and trustee under the will of Edwin Tunis. Used by permission Harper Collins, publisher.)

Southwestern Indians built adobe or stone dwellings where they stored corn in bins or in entire rooms.
(Source: Copyright © 1959 by Edwin Tunis. Revised copyright © by Maryland National Bank, Executor and trustee under the will of Edwin Tunis. Used by permission Harper Collins, publisher.)

constructed of mud, stone, and wood poles. One often entered through the roof — the upper rooms were for living and the lower and back ones were for storage. Corn was kept in enormous clay bins or subterranean. As insurance against drought, the Pueblos kept a reserve of one-third of their crop as a carry-over from year to year. Some Indians south of the Rio Grande, in what is now Mexico, stored grain in 12-foot high vase-shaped containers.

The Pawnee Indians, who lived in an area in what is now Nebraska, considered corn, which they called Mother Maize or Mother Corn, the greatest of all food plants. They believed "she" had mystical powers and even created the first humans from the earth.

Before the World was we were all within the Earth.
Mother Corn caused movement, She gave life. (Novak, p. 338)

In the present day continental United States, Indians grew corn from the Great Lakes to the Gulf of Mexico. Originally corn grew wild and an ear of corn was about the size of a thumb. Early cultivation involved only planting and harvesting. In between those times, Indians hunted, fished, or gathered. Indians found they could increase yields by saving the seed from the better plants to grow new corn. They cultivated the crop more and more and sometimes destroyed the natural habitat of the wild corn. This, along with the destructive forces of worms and other natural enemies, caused corn to lose its power of self-production and it became more dependent on human intervention to grow. As a result, Indians spent more time cultivating and less time hunting, fishing, and gathering. They built larger and sturdier corn cribs and adobe storage bins which resulted in more permanent communities.

In comparison with Europeans, Indians raised food by simple methods, tools, and weapons. Animals ran wild; there was minimum soil tillage. These methods kept the land pristine, but obtaining food was full-time work. The Europeans, in contrast, used plows and draft animals to till the soil while they grazed, fed, and sheltered animals. These methods, while harder on the environment, freed the settlers to explore, to do missionary work and to engage in the arts and sciences.

ETHNIC BARNS HELP COLONISTS SURVIVE: 1607-1776

Stillness Broken by an Alien Tide

If one looked down upon what is now the eastern United States in early 1607, one would have sensed vast stillness. The land east of the Mississippi was wooded with birds singing, leaves rustling, and few human sounds. Small Indian villages along the coast sheltered 40 to 200 people who didn't have noisy muskets, iron axes, mallets, or saws. They had seen English reconnaissance ships sailing up and down the coast. That April the stillness was broken as a little over 100 Englishmen arrived at a peninsula in the Chesapeake Bay in what is now Virginia. They called their settlement Jamestown. The English soon settled other colonies: the Pilgrims at Plymouth in 1620, the Catholics in Maryland in 1633, and later other colonists in New Jersey, the Carolinas, and Georgia.

Other Europeans flocked to America in the 1600s. The arrival of the settlers must have appeared to the Indians as a flood of illegal immigrants. The Dutch settled in New York and later in New Jersey, the Swedes in New Jersey and Delaware, and the French in Quebec and along the Mississippi River. In the West, the Spanish were exploring and settling what is now Texas and New Mexico. Each introduced their own style of barn building.

Agricultural production in the early settlements was essentially nonexistent. Settlers depended on hunting, fishing, trading with the Indians, and supplies from relief ships. People were not familiar with climactic and soil conditions of the new world. They tried communal farming; instead of cooperation there was dissension.

In the fall of 1607, Jamestown settlers suddenly realized they did not have enough food for the winter. They had not cleared enough land or planted crops, and also lacked knowledge about farming. Instead of farmers, these early settlers were seamen, prisoners, gentlemen adventurers, and soldiers of fortune. Men looked for gold rather than attending to the crops. Only one-third of the Jamestown settlers survived the first winter — many starved.

Religious dissenters, not farmers, went to New England. However, New Englanders fared better than the Jamestown settlers because the Indians taught them how to plant squash and corn and to fertilize the crops with fish. Farming in the colonies became a mixture of traditions. The Indians taught the colonists to raise corn, squash, beans, tobacco, pumpkin, and sweet potatoes, while the settlers continued their European traditions of raising wheat, barley, livestock, vegetables, fruits, grass pastures, and hay.

The early barns were crude and temporary in Jamestown, Plymouth, and other colonies. Similar to Indian wigwams, they were constructed of poles stuck in the ground and covered with bark or skins. The settler's priority was to shelter people, not livestock. They fenced their cattle, sheep, chickens, geese, and pigs with woven sassafras saplings or let the animals roam freely on the commons. As conditions improved for humans, the settlers turned their attention to sheltering livestock.

Milk cows and calves, since they were the most valuable and the most vulnerable livestock, were the first animals provided shelter. Little distinction existed between the barn and the house, and animal

This replica of a 1660 tobacco barn at St. Mary's City in southern Maryland has a steep roof and wood shingles.

shelters were often attached to the house. Some early communities used communal granaries and other storehouses to store their crops.

Ethnic and Regional Barns Emerge

Within a few years the settlers were replacing the early crude barns with more substantial ones. Initially the steep roofs on the barns were thatched, but wood shingles soon replaced thatching. The siding was either clapboard or wide vertical boards and was attached to girts with hand-forged nails. Some farm buildings used wattle and daub (small branches or saplings coated with clay) for wall materials. Most barns did not have windows, and if they did, they were small. Barns usually had one or two large doors so the farmer could bring in a wagon or a cart. The type of barn built in any region depended upon the region's agriculture and the builder's ethnic tradition.

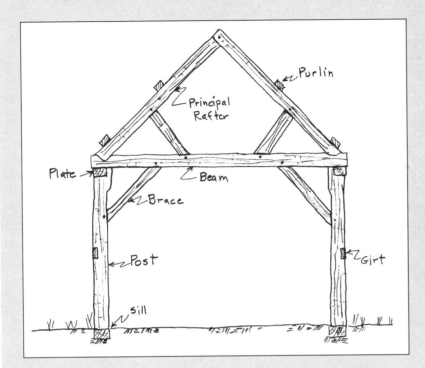

This cross-section of a 17th-century American barn shows the heavy timber frame.

The Barn Builder

The early barn builders were in their late teens or twenties and were 6 or 7 inches smaller in stature than people today. Wood was the main building material; use of stone and brick was limited because of problems getting the lime for mortar. The main barn building tool was the axe. Some builders used a special trimming axe called the adze; the circular saw hadn't been invented. A few saw mills used large hand or water operated whipsaws, but farmers did not have the means to transport the logs to a mill and then to bring the sawed lumber back to the barn site. Many farmers built their own barns; but the larger and better barns were built by hired carpenters or builders. Often the builders exchanged labor with the farmer rather than working for pay.

Barn builders hewed large posts and beams and assembled them with mortise and tendon joints to make a large wood frame or bent. The joints were held together with wood pegs called treenails or trunnels. A barn was built using several of these bents spaced 8 to 16 feet apart. Girts or other secondary framing held the bents together. The sweat and toil that went into the construction are unimaginable. One can't help but admire the precision of the joints achieved with crude tools on such huge timbers.

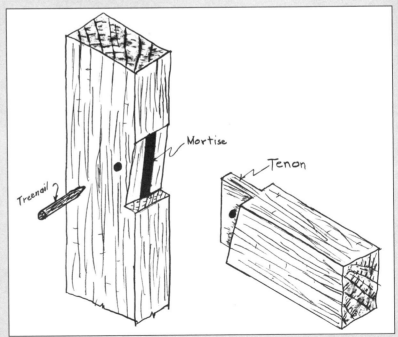

The framing was held with mortise and tenon joints. The tenon or flange fit into the mortise or socket and was held with a wood peg called a treenail or trunnel.

The English barn had a gable roof and large doors on the side for wagon or cart access to the threshing floor. While this drawing shows several windows, most early English barns had few, if any, windows. (Source: Whitaker)

The English Barn

The type of barn built in a particular region depended upon the region's agriculture and the builder's ethnic traditions. In the North the cool rocky soil was best suited for European livestock — cattle, pigs, and wool production. Where possible, the settlers kept the cattle on the coastline on necks and peninsulas and controlled them with minimum fencing. Livestock were first sheltered in northern New England because of the cold, damp weather. The farms were small family units, with most of the land in woods and pastures and only a few acres in cultivated crops or orchards. As farmers grew more wheat and other small grains, they built barns with large threshing floors. As they grew more hay for livestock, the barns were built with larger hay lofts.

The English barn, also called the Yankee, the Connecticut, or the three-bay threshing barn, was simple and plain and was the most common barn built in New England. It provided space for threshing wheat and shel-

tering livestock. The barn was one story and was constructed with four bents that formed a barn three bays long. Barns were either one or two bays wide. The main aisle ran the width of the building and the doors were located in the side of the barn. The center aisle served as a thrashing floor and to one side was a bay for the livestock and the other side was for hay or grain. Sometimes the grain bay would have a hay loft. One unique feature of these barns was that the posts were often flared — called gun stock posts. Farmers probably started building English barns around the middle of the 17th century.

Log Barns

In the middle colonies (southern New York, New Jersey, Delaware, and Pennsylvania), settlers also kept cattle and raised grain, but often didn't shelter the cattle. The Swedes and Finns who settled in New Jersey and Delaware brought log barns to America. There is a restored Swedish Granary at Greenwich, New Jersey, which the local historical society modestly claims "to be the sole surviving example of a farm building erected by the earliest European settlers in our region." One could argue this may in fact be the oldest surviving barn. It is built of cedar log and has a grain floor and area for sheltering farm animals and a loft above.

The log buildings in this replica of a 1638 log farmstead at New Sweden near what is now Bridgeton, New Jersey, include a residence, a blacksmith shop, and several barns. The building on the right in the back is a threshing barn. Today a Swedish granary stands in nearby Greenwich, New Jersey. (Source: New Sweden Farmstead Museum, Belva Prycl, Artist)

This restored Swedish Granary, built circa 1650 and which stands at Greenwich, New Jersey, may be the oldest surviving barn in America. (Source: Cumberland County Dept. of Planning and Development)

But the Germans, and not the Swedes or Finns, are primarily credited with spreading the log barns throughout the mid-Atlantic colonies and southern Appalachia. In contrast to post and beam framing, the timbers were laid horizontal on top of each other and joined at the corners with various types of notching. (For more information on log barns, see page 52.)

The Dutch Barn

The Dutch barn was based on a prototype found in the Netherlands and was built mostly in New Jersey and New York probably starting around 1680, if not earlier. The Dutch first came to America as settlers to the New Netherlands in New York, they did not come as immigrants to a British colony. As a result, there was probably less exchange of barn building techniques between the Dutch and the other colonists. The Dutch barn had three aisles like many European churches; the center one was a wide threshing aisle and there were two narrower side aisles. One side aisle was devoted to cows and the other for horses and grain storage. In contrast to the English barn, the aisles ran the length of the building. The doors were in the gable end walls. Dutch barns had steep gable roofs and

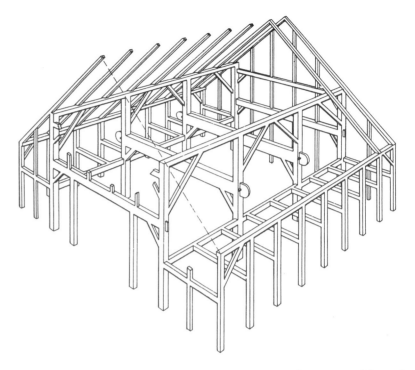

Dutch barns had massive anchor beams spanning the center aisle and they were connected to the post with a unique mortise and tenon joint. The ends of the beams had a tongue-shaped tenon that extended past the post. (Source: Jack Sobon, Windsor, Massachusetts)

Dutch barns built in eastern New York had massive gable roofs, low side walls, and doors in the gable end. (Source: Whitaker)

low side walls. The main structural frames were "H" shaped. A large anchor beam formed the horizontal part of the frame. These anchor beams were as thick as 18 inches and weighed as much as a ton. They were connected to the columns (the vertical parts of the "H") with a unique mortise and tenon connection. The Dutch sometimes used shingles instead of clapboard siding. The large sweeping roof and low side walls gave it an elegant appearance.

Swiss German Pennsylvania Barn

The Germans had a particularly good reputation as farmers; they used manure on their fields, rotated crops, stayed where they settled, and built substantial barns. The English and the Scotch, on the other hand, tended to be careless with the soil, to build more temporary barns, and to move when the soil was depleted. The Germans, along with the Swiss, started building barns around 1700 in southeastern Pennsylvania; their earliest attempts were single story log buildings. Later they built the more typical Swiss German Pennsylvania barn. It had two stories with livestock kept below. A second floor overhung the first floor with a cantilevered forebay. On the second floor there were three bays — a center one for threshing and two side bays — one for hay, the other for straw. The barns were built with earthen banks or ramps which enabled farmers to drive their wagons onto the second floor. As a result, people started calling these buildings "bank barns." Soon these barns spread to other parts of Pennsylvania, to Maryland, and into Virginia's Shenandoah Valley. (For more information on the German Pennsylvania barns, see page 57.)

Germans, for the most part, built larger and studier barns than the English and typically they were built into a bank with a second floor forebay that extended past the first floor. (Source: Whitaker)

French Barns

The French were fur traders and explorers along the Mississippi River and the Great Lakes. They developed few permanent farming communities. When they did build farm structures in the Mississippi Valley, they used a unique system of sinking posts several feet into the ground — called "poteaux-en-terre" (post-in-the-ground). These vertical posts had spaces of a few inches between them and these were filled with a mixture of clay and grass, or clay and Spanish moss in the deep South. Often the roof overhung the main structure to form an aisle or "galerie" on three or four sides of the building. For the most part the French did not build horizontal log buildings. The exception was in Wisconsin and the upper Great Lakes where they built log barns called the "piece sur piece" (piece on top of piece).

Along the St. Lawrence River and the Great Lakes, the French built the Quebec long barn. These heavy timber barns were similar to the English barn except they were long and narrow. The walls were covered with vertical planks or were made of wattle and dab construction. The barns were elongated because several farm buildings, such as a residence for a tenement farmer, were included with the barn and were all connected together under one roof. This arrangement enabled the farmer to do his chores in the severe winters without venturing outside.

Tobacco and Other Southern Barns

Tobacco dominated agriculture in the Chesapeake Bay area and as a result small tobacco curing barns dotted the landscape. Tobacco barns were built with light wood framing closely spaced together. It served as structural support for the roof, as well as scaffolding to carry the sticks holding the tobacco leaves.

Three types of tobacco barns were eventually built in America depending on the three methods of curing tobacco: fire, flue, or air. In fire-cured barns, small slow-burning hardwood fires were built on the barn's earth floor or in shallow trenches in the floor. The rising smoke turned the tobacco leaves dark brown and added flavor to the tobacco curing on racks above the fire. These barns tended to be rectangular, tall in height, and air tight.

In flue curing, fires were made in furnaces or kilns under sheds adjacent to the barn. The heat from the fire was carried throughout the barn through long ducts or flues and cured the leaves a light yellow. The smoke did not come in contact with the tobacco. These barns tended to be square and tall but often smaller than the fire-cured barn. The distinguishing feature of the flue-cured barns were sheds or overhangs on the side of the barns. These were used to shade the workers who prepared the leaves for hanging in the barn as well as protecting the furnace and fuel from the weather.

In air curing, the tobacco was allowed to dry in well-ventilated barns. These barns came in a variety of shapes and sizes, but tended to be larger than either the fired or flue-cured barns and were often long and rectangular in shape. Air-cured tobacco was grown in a variety of regions including areas outside the South, such as Pennsylvania and Connecticut. Ventilation was often provided with vertical wall panels hinged at the top. Other ventilation methods included horizontal wall panels, opening under the eaves and roof ridge ventilators.

South of the Chesapeake Bay region, in the Carolinas and Georgia, rice and indigo were the chief crops. In the Carolinas they "raised" cattle which were neither fed nor sheltered but turned loose to roam the woods and fields. Even grain threshing was often done outdoors. Sometimes small crib-type structures were built for storing corn and housing select livestock. Few heavy timber frame barns were built in the South because the barns were small and the roof framing didn't have to support heavy snow loads.

Tobacco and rice were very labor intensive, and this led to the plantation system where slaves and indentured servants were used. Plantations often had thousands of acres and elaborate farm buildings. For every plantation there was a score of small subsistence farms where the farmers built temporary structures and moved when tobacco had depleted the soil.

This tobacco barn, originally probably a fire-cured barn and later converted to a flue cured, is one of the many early American farm buildings one can peruse at the Museum of American Frontier Culture, Staunton, Virginia.

Southwestern Barns

In the Southwest, the Spanish organized farming among the Indians. They encouraged the Indians to raise cattle and introduced wheat, barley, watermelon, grapes, and other fruits as well as goats, donkeys, horses, and mules. This region was one of the few where the Indians adapted European agriculture.

In the early 17th century, the Spanish erected log shelters and simple huts called jacales made of poles covered with branches, grass, reeds, and mud. These huts were first used as human shelters but as the settlers built larger and more permanent structures, these huts were used to shelter the animals. By the 1700s the Spanish were building permanent adobe and stone structures such as missions, presidios, forts, and haciendas. Grain storerooms and corrals, where they kept horses and prize cattle, were built as part of the main buildings. All the buildings were built together to form a fortress to protect them from bandits and raiding Indians.

End of the Colonial Period

The British colonies continued to gain dominance over the French and the Spanish in the 1700s. At the Treaty of Paris in 1763, the British gained Canada, the land east of the Mississippi and Florida. The conflict continued to grow, however, between the Colonies and the Crown, and on April 19, 1775, the American Revolution began.

By 1775 the American colonies that had started with about 100 settlers at Jamestown had grown to 2.5 million Europeans. They had settled within 200 miles of the Atlantic Coast. Most of these immigrants were young; fertility rates were high; and the population burgeoning. In the South, the labor-intensive crops of tobacco and rice created a demand for workers. The Dutch brought a boatload of Africans to Virginia in 1619, and by 1775 there were about one-half million Africans in America; most were slaves in the South. Fewer Indians lived in this area than when the Europeans first arrived. Many Indians along the Atlantic seaboard had been driven out by the Europeans and many died of white man's diseases such as smallpox, measles, and typhus.

The people in 1775 were still farming much the same way as the first settlers, and for the most part in the same way people had farmed for thousands of years earlier. They planted crops by hand, weeded with a hoe, and cut grain with a scythe. Only a few farmers had oxen, horses, or plows, and those who did performed custom plowing for others. Few new agricultural tools were developed during the colonial period. However, farmers learned to use better plant varieties and livestock breeds, to store crops effectively, to find the most productive land, and to shelter their livestock. Cross-cultural transfer of information between the various ethnic groups and a commercial mentality created a strong agriculture. While many of the first settlers died of starvation, the food supply improved rapidly. By just their second year in America, the Pilgrims were celebrating Thanksgiving because of their abundant corn harvest. Many early settlers continued to suffer food shortages and hardships, but by the middle of the 17th century most farmers were meeting their own food needs. And by 1775 they raised enough food to trade and sell to the artisans and craftsmen in the colonies. They even produced enough food to trade overseas. Farmers needed good barns to support this growing agricultural production.

During the 1700s, professional architects began designing the bigger homes, schools, churches, industrial buildings, and government buildings. Architectural books were published. The newer buildings used more brick, had larger windows, used more trim, molding, and paneling. These new designs, called Georgian architecture, found their way into some southern plantations. However, farmers and carpenters, not architects, were still designing barns, stables, and carriage houses. The distinction between shelters for people and for animals was getting greater. Barns were simple and crude compared to the ornate and sophisticated architecturally-designed buildings.

The stillness of the North American continent had broken nearly 170 years earlier. It started with below subsistence agriculture, settlers struggling just to keep from starving, and hovel-like barns. These conditions were now replaced by bustling economies, thriving seaports, and abundant trade. Tobacco, livestock, rice, and indigo moved out as exports, and slaves, tools, building materials, and textiles were imported. The strong agricultural economy demanded larger and better barns.

THE NEW NATION MOVES WEST: 1776-1860

After the Revolutionary War, most people of European extraction were still farmers and lived along the East Coast. Many of those who lived in town had their work tied to agriculture — blacksmiths, millers, and cotton gin operators. Boston, New York, Philadelphia, Baltimore, and Charleston were the only cities of any size. The farmers were building the English, the German, and the Dutch barns, and small log cribs on their farms. Indians, who were being driven westward, continued their old methods of obtaining and storing food. A few Indians in the Southeast were adopting the European agricultural practices.

A tremendous pent-up demand for western land existed after the Revolutionary War. Before the war, George III forbid the colonists from moving West. Now the colonists had not only gained their independence, but also access to vast new lands between Appalachia and the Mississippi River. The new stronger central government, established by the Constitution, promoted western growth. Passing the Land Ordinance of 1785 provided for government land surveys — a key factor in the orderly settlement of the West. Improvements in transportation opened the West. The National Road, also called the Cumberland Road, was started in 1811. This road eventually provided a link between Baltimore and Vandalia, Illinois. After the War of 1812, thousands of people poured over this road to settle in Ohio, Indiana, and Illinois. Pioneers brought their cows with them tied to the wagons. The cows provided manure, milk, and meat. As the settlers moved West, they also brought their barn building tradition. However, farmers often had to adapt their traditional building methods to meet the new conditions of the new land.

Regional Barns in an Expanding Country

Settlers moving into Appalachia grew corn and raised cattle and pigs. The wooded hilly land often allowed only subsistence agriculture. At first farmers didn't build substantial barns. They let their pigs and cattle run through the woods in the early 1800s or kept them in small log cribs. It wasn't until the middle of the 1800s, when these frontier communities became more established and prosperous, that substantial English and Pennsylvania bank type barns were constructed.

New Midwestern farmers also primarily used the English and Pennsylvania bank barn designs. However, some farmers on the prairies of Illinois and Iowa found there were not enough large trees to build their traditional barns. To protect cattle from the cutting prairie winds and driving rains and snow, they built flimsy structures with pole framing that supported a thatched coarse hay roof. The shelters were a cross between a barn and a haystack. They were open on the sides or may have had hay racks along the windward side. Over the winter the cattle ate these shelters (*The Prairie Farmer*, March 14, 1861).

In the Midwest some farmers built temporary barns with pole frames and thatched roofs. (Source: Bryan David Halsted)

In the far West, Americans came from the east, Spanish and Mexicans came from the south, and Russians came from the north. Most of the far West was not even a part of the United States until the 1848 War with Mexico. In the Southwest it was a different culture; Spanish was the main language. The Spanish and Mexicans continued building missions, forts, presidios, and haciendas; they

Las Golondrinas, a restored major 18th-century ranch, is located near Santa Fe, New Mexico. The structure toward the back on the left is a shelter for the farm animals.

Agricultural Changes

A spirit of agricultural improvement dominated the early 19th century. Agriculture had been based on superstition, astrology, and tradition. For example, farmers feared that the iron plow poisoned the soil, but superstition was being replaced by scientific studies and innovations. New machinery was invented and the fertilizer industry was born. Agricultural societies were founded promoting better ways of farming (one of the first was The Philadelphia Society for Promoting Agriculture). These societies distributed agricultural information on programs and methods that had been developed in other countries. Scientific papers were published in new agricultural periodicals such as *The American Farmer* founded in 1819.

Factory-produced farm machinery changed agricultural practices including how barns were designed and built. Before the Revolutionary War, most plows and other farm machinery were forged in local blacksmith shops. From 1820 onward, there were continual improvements with the standardization of the plow, horse-drawn cultivators, grain drills, and corn planters. In 1837 John Deere developed his first plow with steel shares particularly suited for prairie soils. The new machinery enabled farmers to produce more. Farms grew in size and farmers needed larger and more substantial barns to store their larger harvests and to house their growing herds.

Harvesting had always been the bottleneck that limited the amount of cereal grains that could be grown. In the first half of the 19th century, most farmers harvested their wheat, barley, and oats

were more elegant and permanent than previous ones and were built of adobe, masonry, stone, and some had red clay tile roofs. Longhorn cattle were an important part of the agriculture along with wheat, sheep, and barley. Ranching operations were closely associated with these facilities. Stables, tack rooms, hay storage, and granaries were built as an integral part of the ranch house. (For more information on adobe ranch buildings, see page 62.) This culture and their architectural style started vanishing when Americans and Europeans moved into the Southwest in 1821, lured by cheap land and gold. They introduced sawed lumber and different building designs. In 1833 the Mexican secularization laws ended the church's construction of new missions and some were closed or fell into shambles. In 1848, Mexico conceded most of their holdings in the Southwest to the United States.

Another Revolution

The American Revolution was followed by another revolution — The Industrial Revolution — which promoted mass production. Industry was particularly strong in the North where there was ample water power to operate mills and factories. The nail cutting machine developed in 1777 and the power circular saw developed in 1814 started to slowly move barn construction from heavy hand-hewn timbers connected by wood pegs to light-framed nailed lumber.

The development of Cyrus McCormick's reaper in 1831 was a major factor in ending the eastern threshing barn.
(Source: Photo courtesy of the State Historical Society of Wisconsin)

with a hand cradle scythe — a very slow process. From 1820 through the 1860s, enormous strides were made in machinery for harvesting, threshing, and winnowing grain. In 1831 Cyrus McCormick put his horse-drawn reaper on the market. It was ideally suited for the flat prairies of the Midwest. Now Midwestern farmers could harvest large amounts of grain and Northeastern farmers couldn't compete with them. Threshing and winnowing was moving from a barn operation to a field operation. Farmers were no longer building threshing barns.

The East

In the East, Northeastern farms remained small scale and often inefficient. Around 1800, the typical Northeastern farm had less than 120 acres and perhaps only about 15% of the land was planted. The rest was in pastures or woodland pastures. After 1840, Northeastern farmers could not compete with Western farmers in wheat production, so they moved into vegetable, hay, milk, butter, cheese, and fruit production. Northeastern farmers, particularly those in New England, started housing more of their livestock during the winter because they found that sheltered cattle produced more. Farmers modified their old threshing barns into general purpose livestock shelters. They built bigger and more substantial barns. Near the close of the Revolutionary War, the typical English barn was about 20 by 30 feet, and 40 years later they had grown to about 26 by 40 feet.

This change to general purpose farming encouraged Northeastern farmers to replace the English barn with the more flexible arrangement of the New England barn. This barn had one story plus a loft and was usually three bays wide. The main door was in the gable end and opened into an aisle running the length of the

The New England barn, with the door in the gable end wall, often had a transom over this doorway and a large cupola on the roof. Early New England barns didn't have basements as this barn in Franconia, New Hampshire. (Source: Library of Congress)

building. These barns were usually three or four bays long, but they could be easily expanded by adding more bays to the length. New England barns were also built of hewed heavy timber, but after 1840 more sawed lumber began to appear. After 1850, many barns, particularly in the East, had cellars or partial cellars similar to the Pennsylvania bank barn. The English and New England barns were often part of a farmstead system where the buildings were all connected. The main house was connected to the barn through a series of buildings — the kitchen, the workroom, and the wagon shed. This enabled the farmer to move freely from building to building in the winter despite the bitter cold and drifting snow. Connected barns were mostly limited to the six New England states and New York where the winters were the most severe.

Farms in the mid-Atlantic states were larger than the farms in New England. New England farms were limited in size by the rocky, hilly terrain. For a while Pennsylvania was a leading wheat producing state. Farmers in Maryland and Virginia also moved to wheat because tobacco had worn out the soil. Most mid-Atlantic farmers let their cattle roam and did not house them. The exception was the Germans who felt it was cost-effective to house their cattle. By housing cattle they could control feeding and breeding and reduce illness.

At the time of the Revolutionary War, the Germans were building Pennsylvania bank barns. These barns, described earlier, were two-story buildings with an earth ramp or bank serving the second floor. The second floor extended beyond the first floor and the extension was call a forebay. After 1790, farmers starting replacing these early small German Pennsylvania barns with the standard Pennsylvania bank barn. The newer barns had a lower pitched roof, shorter fore-bays, and were larger than the earlier barns. They were as long as 100 feet. Between the end of the American Revolution and 1860, the Pennsylvania bank barn moved into Ohio, Indiana, and Illinois, generally following the path of the National Road. It also moved into Wisconsin and Michigan. (For more information on the Pennsylvania bank barn, see page 57.)

By 1776 Dutch barns, not to be confused with the German bank barn, were found in eastern New York, New Jersey, eastern Pennsylvania, southwest Connecticut, and Long Island. After the Revolutionary War, the Dutch colonies disbanded and fewer Dutch barns were built. Around 1850 Dutch immigrants settled in southwest Michigan and built barns similar to the Dutch barns of New York.

The mule stable on America's most famous plantation, Mount Vernon, is located about 15 miles south of Washington, DC. Besides the main mansion, there were about 20 outbuildings some of which stored farm machinery and kept animals.

The South

In the South, the most rural area of the new nation, labor-intensive crops like tobacco promoted gang slave labor and large plantations over 5,000 acres. Mount Vernon, George Washington's estate, had five farms and 8,000 acres. Washington first grew tobacco as his major crop and later replaced it with wheat. Like other Southern farmers, he moved from building tobacco barns to building threshing barns. Across the South, tobacco was over-produced and ruined the soil for further tobacco growing; rice had been overproduced; and the English indigo market had collapsed. In most areas, cotton took their place aided by Eli Whitney's cotton gin, patented in 1794. Labor-intensive cotton increased the importance of the large plantation, often described as a "beautiful but cruel work place." During this period, plantations spread into Alabama and Mississippi and along the Mississippi River. By 1860, there were more than 46,000 plantations in the South — many with more than 20 slaves. However, there continued to be many subsistence farmers with fewer than 100 acres.

There was no typical Southern barn; they varied with the climate and the ethnicity and wealth of the owner. Frequently farmers didn't house cattle or cows and often milked the cows in open corrals. Few new tobacco barns were built in the old tobacco producing areas, but small crib barns continued to be built. At the other extreme were the elegant plantation barns. While there were a few large multi-purpose barns, Southerners tended to build specialized separate buildings: carriage houses, kitchens, smoke houses, tool rooms, pig houses, mule stables, and horse stables. Fine riding horses were important plantation status symbols and owners built stylish stables. The larger plantations had rice mills, cotton gins, and presses; the plantations along the Gulf Coast had sugar mills. Because of the French population living there, elegant barns in southern Louisiana had hip roofs and extensive eaves resembling barns from northern France.

The Amana Church Society started a communal community near Iowa City in 1855. Like other communal groups they built large farm buildings such as this corn crib.

Barn Fads

From social abuses of the Industrial Revolution and the plantation system, arose social reformers crying for abolition, better labor laws, equitable land distribution, and women's rights; and utopian communal societies grew, particularly from 1820 to 1850. These societies were primarily religious, such as the Shakers, but a few were secular such as the LLano Colony in California. In building their communities, they looked to the Bible, utopian writers, other utopian groups, or direct inspiration from God to their leaders for their architectural designs.

Several elegant barns came out of the utopian movement — one was the Shaker stone round barn built in 1826 in Hancock, Massachusetts. (For more information on round barns, see page 68.) The Shaker leaders believed the barn's design was inspired by God. The barn, which was rebuilt after a fire, still stands and visitors often mistake it for a place of worship. In 1860 the Shakers built another unique barn in New Lebanon, New York. It was five stories tall, built of stone, and was about 300 feet long. Large barns were characteristic of communal societies. The Amana Church Society built large corn cribs in their settlements near Iowa City.

Another movement that influenced barn designs was the growth of new popular architectural forms. From 1820 to 1860 the influence was Greek revival architecture. Libraries, museums, schools, and government buildings were designed to resemble ancient Greek temples with large columns and flat or low pitched roofs. One Southern plantation barn was built in Fluvanna County, Virginia, with large stone columns on the front, but for the most part, characteristics of Greek revival barns were a low pitched roof with trim and horizontal molding that emphasized the triangular shape of the gable end above the barn door. This characteristic found its way into barns of the Northeast.

In the first half of the 19th century, Americans became fascinated with kings, queens, knights, and the places they lived and worshiped — castles and cathedrals. This fascination promoted an interest in Gothic architecture which emphasized height and the

This elegant sandstone barn with columns and Greek pediments shows Greek architectural style, popular in the mid-1800s. This barn was built in 1817 in Fluvanna County, Virginia. (Source: Library of Congress)

vertical as in the great European gothic cathedrals. Elements of this architecture found its way into barns. The characteristics were steep pitched roofs, vertical lines created by board and batten vertical siding, long narrow windows, and large cupolas — things that directed the eye upward.

END OF ETHNIC BARNS: 1860-1920
The Great Plains

In 1860, when Abraham Lincoln was president, farmers were still building heavy timber ethnic barns, particularly the English and the German bank barns. Only a few settlements were west of Kansas City, except for those in the valleys along the Pacific Coast and the Mormons in Utah. On April 12, 1861, the Civil War started. While

This restored coach house in Stuart, Iowa, has many Gothic features emphasizing the vertical as found in European castles and cathedrals.

Limestone barn built by Czech immigrants near Dorchester, Nebraska.

the war raged in the East, settlers pushed on to the Western Great Plains although many considered the area uninhabitable. After the war, settlement in the West increased at an even greater pace. The pioneers first settled along rivers and streams where there was water and wood, and later moved onto the open treeless plains.

Several factors encouraged western settlements: the 1862 Homestead Act giving prospective farmers 160 acres of land; development of plows that could break up the prairie soils; the removal of hostile Indians; and the expansion of railroad lines. Ethno-religious societies, such as the Irish-Catholic Colonization Society, the Swedish Agricultural Company, and the Hebrew Union Agricultural Society helped locate land for prospective settlers. The railroads recruited immigrants from northern Europe and eastern United States. Throughout the Plains various ethnic groups settled. The Kastanek limestone barn near Dorchester, Nebraska has a stone set in the wall with words chiseled in Czech about the barn's history. This barn may not strictly represent a Czech ethnic barn, but it does illustrate the good stone masonry work typical of many central European barn builders.

Cattle production expanded into the Western Great Plains in the 1860s with much of the cattle coming from ranches formerly owned by the Mexicans and Spanish. Grain Plains ranching didn't require barns to house cattle but demanded only small out-buildings for horses, saddles, and maybe chickens. This was the period of the great cattle drives, where cattle were driven to Abilene, Wichita, Topeka, and Dodge City. By 1880 cattle were big business, but the year also marked the beginning of the end of the cattle drives and the open range. The cattle industry suffered from low prices, and sheep herders and farmers moved in, closing the open ranges with barbed-wire fencing.

The treeless barren plains offered few building materials for the new farmers and herders. They built sod structures from 1860 to 1900, primarily houses, but in some cases shelters for milk cows, calves, and chickens. "Soddies" were built throughout the Great Plains. Turf was cut by a grasshopper plow into sections called Kansas bricks, and laid up with staggered joints. The walls were 1 1/2 to 3 feet thick and the roofs were made of pole rafters and covered with brush, hay, prairie grass, tar paper, or shingles. The Mormons were credited with first employing sod construction in

As farmers built better dwellings they used their sod houses, like this Central Great Plains sod house, for barns. (Source: Museum of the Cherokee Strip)

The barn owned by Elijah Filley was built of limestone in 1874 in Gage County, Nebraska.

This map shows the distribution of sod buildings. (Source: Noble, Vol. 1)

America, but Russian-German farmers had used this technique in Russia. While these houses and barns provided good insulation, they were dusty and muddy.

Where it was available, stone was used on the Great Plains. Individual and clusters of stone barns were built only where limestone, sandstone, or field stone

occurred, for example in the Flint Hills of Kansas. A few farmers and ranchers built grand stone barns on the Plains between 1870 and 1910. A most elegant example is the Filley stone barn built in Gage County, Nebraska, in 1874. It was built from an outcropping of limestone on the farm, with walls 24 inches thick and beautiful arched windows and doors. The Kimmell barn near Covington, Oklahoma, was built of sandstone and has large fine arched doorways made of hand-chiseled stone.

The Kimmell stone barn was built in 1906 in Garfield County in north central Oklahoma. Behind these wooden doors painted with an arch is a real stone arch. See photo on page 134.

Oregon Territory/The Northwest

The Oregon Trail promoted growth of the Oregon Territory in the Northwest. Created in 1848, it included the states of Idaho, Washington, Oregon, and parts of Wyoming and Montana. The trail carried 12,000 settlers in the 1840s. It was the longest overland trail, running 2,000 miles from Independence, Missouri, to Fort Vancouver. It took settlers a hard six months to make the journey. In 1883 the railroad was extended into the Pacific Northwest and the trip took less than a week by rail. The new settlers brought their eastern traditions of farming and barn building to the territory. West of the Cascade Mountains, settlers first built log barns. Soon they were replaced with the English barn or forebay barns similar to the Pennsylvania bank barn. The early Northwestern barns were general purpose barns with a threshing floor, grain bins, hay mows, and stalls for horses and cows. After 1890 more barns were built that were designed primarily for dairy cows.

In the Columbia Basin and eastward, the settlers produced wheat and cattle; the feeder barn (described on page 22) was characteristic of this area. Lean-tos for loose livestock were often constructed on one or more sides of the central barn area where hay was stored. The roof pitch of the lean-to was often the same as the roof pitch of the main structure. This gave the barn a long sweeping roof line somewhat similar to the lines of the early Dutch barns.

Later, as barn designs became more standardized and promoted nationally, Northwestern farmers, like farmers elsewhere, were influenced by what was popular at the time. They built the Wisconsin dairy barn and even built a few round barns.

The Southwest

In the Southwest, California moved from intensive wheat and cattle production into fruits and vegetables. Farmers and ranchers found that they could get high yields and better income by growing fruit and vegetables with irrigation. By 1910 there were many large-scale commercial fruit and vegetable farms in California. Southwestern farmers followed a similar pattern of bringing their eastern tradition of barn building. However, most livestock were kept in open corrals with small sheds for sheltering them from the blazing sun. Ranchers built large horse barns with a high center section and stalls on each side in lower shed-like structures. The barns had wide doors at each end and vents under the roof eaves of the center section. This arrangement provided good ventilation which gave relief to the horses from the hot dry summers of the Southwest and helped control odors.

This cattle barn, with the large sweeping gable roof near Glenwood, Washington, is representative of some Pacific Northwestern barns built at the turn of the 20th century. (Source: M.D. Locke Jr.)

in eastern Tennessee. These barns had a double crib covered with an oversized loft area that was cantilevered over the cribs. The cantilevered loft provided a large overhang on all sides of the cribs (See the Tennessee cantilevered and other log barns on page 53). Southern farmers remained poor until the New Deal and World War II. Consequently, few farmers were able to afford to build new or well-built barns.

The North

In the North the number of farms declined because of their inability to compete with Western farmers and many of the farmers that remained turned to dairy. A surprising number continued erecting the old-fashioned post and beam hand-hewn timber barns late in the 19th century. During and after World War I, some farmers built the Wisconsin dairy barn developed by the University of Wisconsin. The barn soon spread through the dairy regions of the Great Lakes, New York, and New England. These barns were about 36 feet wide and about

The Southeast

The Southeast struggled to recover from the Civil War. The war left the barns, houses, and fences in ruins. Seventy-five percent of the farmers were share croppers or tenants; many remained in constant debt; plantations were sold; and a few freed slaves acquired small farms. Southern farmers built wood-frame corn cribs, smoke houses, feeder barns, and small general purpose barns. In Appalachia they continued to build log barns. Tennessee cantilevered, double-crib log barns were built in the latter part of the 19th and early 20th centuries in southern Appalachia — mostly

In the early 20th century, Appalachian farmers were still building the Tennessee double crib cantilever barn. (Source: Noble. Vol. II)

This heavy post and beam framing is in a 19th-century barn near Diane, New York. (Source: Bruce Brittain)

About the time of World War I, the Wisconsin Dairy barn was built in New England and around the Great Lakes. (Source: Noble, Vol. II)

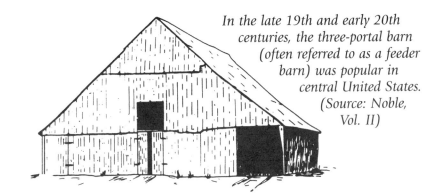

In the late 19th and early 20th centuries, the three-portal barn (often referred to as a feeder barn) was popular in central United States. (Source: Noble, Vol. II)

100 feet long and had two rows of stanchions running the length of the building. Windows along the long walls provided good lighting and ventilation. Builders usually made the barns of lumber and with truss gambrel or Gothic roofs.

Silos started to be an integral part of the dairy barn after 1875. They made it convenient to move the silage from the silo to the cattle feeding troughs in the barn. While the University of Illinois first experimented with silos, most of the first ones were built in New England and New York. The first silos were pits, next rectangular wood towers were built, and finally circular wood stave or masonry silos were used. The silo reduced space requirements for hay lofts because farmers fed more silage and less hay. The silo changed the farmstead landscape by adding these upright vertical structures to the barn.

The Midwest

Midwest farmers continued to raise corn and hogs and many built the German Pennsylvania barn in the 1800s. These barns became larger and were now built with extended forebays up to 20 feet long supported by columns instead of being cantilevered. The Pennsylvania barn continued to spread beyond the Midwest as even some were built in California, Oregon, and Washington.

A new barn was emerging in the Midwest in the late 1800s. Midwestern farmers were purchasing range cattle from the central Great Plains, fattening them, and then shipping them to Chicago and other large cities. They needed barns to protect the cattle in the winter and to store hay and grain. The new barns were called feeder barns. They usually had four rows of pens and three aisles. Doors at the end of each aisle gave rise to another name — the three-portal barn.

The three-portal barn, with doors at the gable end, had huge roofs and usually had a large center hay door protected with a hood. They did not have basements. Some were built with a single-pitched gabled roof covering all the pens. Others had a steep-pitched gabled roof over the two center pens and a lower pitched roof over the outside pens. Sometimes the gable ends of the barn were longer than the sides.

By the late 1800s, large timbers were becoming scarce in the Midwest and farmers were looking for open hay lofts without cross beams that obstructed hay carriers. Lighter sawn lumber was already being used in house construction. Barn builders started using plank construction, an intermediate between heavy timber framing and the light lumber framing used today. In plank construction, carpenters used 2-inch thick boards of varying width. Often they were doubled and tripled to make bigger members. Nails and spikes replaced the mortise and tendon joints. Instead of making heavy square beams, they made narrow and deep ones, which were more efficient for resisting bending. Roof trusses, which provided open lofts, replaced roof bents and cross framing.

Round Barns. Round and octagonal barns became a fad primarily in the Midwest, although they were built throughout the United States in the late 19th and early 20th centuries. While they represent a unique type, they are only a fraction of a percentage of all barns built. Orson Squire Fowler, a mid-19th century reformer and eccentric, is credited with promoting octagonal buildings. He thought octagonal structures improved health and created a better life. He claimed they were tighter and warmer — the room arrangements in an octagonal house would be "handy and convenient" while an unhandy house would create exhaustion and sickness for the housewife. While the round barn, in theory, used less material, it was often difficult to secure experienced carpenters that could build a good round or octagonal barn (For more details on round barns, see page 68).

Victorian Barns. In the late 19th century people became preoccupied with Victorian architecture. Buildings were designed with intrinsic gingerbread details. They have been described as showy and overdressed. Prosperity, along with the development of the

lathe and scroll saw, gave farmers the luxury and means of incorporating these details in their barns. Gentlemen farmers and affluent suburbanites built elaborate carriage houses — buildings for keeping the carriage, horse, feed, hay, and sometimes the stable boy. The carriage house was the auto garage of yesteryear.

Changing Agriculture/Changing Barns

New economic and political forces were revolutionizing agriculture and barn building. After the Civil War, agriculture faced serious economic challenges. The plantation system in the South had been dissolved and many barns, fences, and other farm structures had been destroyed and had to be replaced. As agriculture recovered, it faced new problems. More and more of the items needed by farmers such as fertilizers, seeds, and machinery, came from commercial enterprises outside the farm. Farmers were also purchasing more of the barn building materials — lumber, nails, and hardware — from commercial firms. They were financing their operations from larger and more impersonal banking institutions. More of their farm products were being shipped across the country on a national system of railroads. The individual farm was becoming an integral part of a nationwide economic system. However, many farmers felt they were losing control and were suffering abuses from this large impersonal system. In response, farmers organized the National Grange and Farmer's Alliances which promoted cooperatives. They fought suspected abuses such as railroad tariffs and sought currency and banking reforms favorable to agriculture. Before these cooperatives built their own community meeting halls, they often met in barns. This new agriculture promoted bigger farms and more sophisticated barns.

New technology was another force changing agriculture and farm building designs. From the middle of the 19th century to the early 20th century, agriculture moved from largely horse power, to the steam engine, and then to the internal combustion engine. A horse powered agriculture had required large barns for the many horses needed to pull the plows and operate the threshing machines. With the conversion to steam and the internal combus-

This Arcadia, California redwood and cedar coach house built at the turn of the century by E.J. Baldwin, a wealthy businessman, shows the popular Victorian gingerbread of the times.

Rural Free Delivery (RFD) started in the late 1800s increasing the circulation of farm journals to all parts of the country. These magazines often carried recommended barn designs. This, along with the fact that standardized lumber could be shipped by rail throughout the nation, helped standardize barn designs. While there were still regional differences in barns, the ethnic style barns like the English, the German Pennsylvania, and the Spanish adobe were disappearing. RFD helped create thriving mail order businesses, mainly Montgomery Ward and Sears Roebuck & Company. Farmers could now order hardware and other accessories for the barn: metal ventilators, tracks for sliding barn doors, and cattle stanchions.

Sears, Roebuck & Company sold barn accessories through their catalog
that helped standardize barns throughout the country.
(Source: Courtesy of Sears, Roebuck & Company)

Galvanized Steel Ventilators

Extra heavy, 22, 24 and 26-gauge galvanized steel is used in the manufacture of our Majestic Ventilators, according to size.

Majestic Ventilators are equipped with four stay rods which hold them securely in place on the roof. A wire screen is provided to keep birds out.

Your choice of gold bronzed horse or cow weather vane.

Prices of Majestic Ventilators for large buildings:

Catalog No.	Size Flue	Size Base	Height	Wt., Lbs.	No. Cattle	
48K657	18 in.	26 in.	7 ft. 6 in.	115	4	$22.00
48K652	20 in.	29 in.	8 ft. 8 in.	155	6	25.60
48K653	24 in.	36 in.	9 ft. 5 in.	182	8	29.20
48K654	28 in.	40 in.	10 ft. 4 in.	235	12	34.95
48K655	30 in.	44 in.	10 ft. 6 in.	252	14	37.35
48K656	36 in.	55 in.	11 ft. 3 in.	396	20	45.60

tion engine, these barns were often converted to cattle barns and eventually were used for calving, hospitalizing sick cows, and for artificial insemination.

Improvements in all aspects of agriculture continued in the late 1800s. There were improvements in the plow, grain drill, planter, reaper, wire binder, threshing machine, and corn husker. Grain threshing continued to move from the barn to the field, thus eliminating the need for the barn's threshing floor. Besides mechanical improvements, there were also chemical and biological improvements in soil analysis, plant pathology, and the discovery that nitrogen could be obtained from leguminous plants. The use of lime, fertilizer, and manure increased during the early 1900s. In 1880 Louis Pasteur published his germ theory, which led to emphasis on better lighting, ventilation, and sanitation in animal husbandry. The dark, poorly ventilated basements of the Pennsylvania bank and other basement barns did not meet these new considerations and few basements were built in barns after the early 1900s.

After 1897, the agricultural economy began to improve leading to the "Golden Age of Agriculture" (1910-1914). Prices were high and stable, and the country was at peace. Farmers built bigger, more elaborate, and finer barns.

THE RISE OF FOOD FACTORIES: 1920-2000
The Depression Years

In 1920 American farmers were prospering. They had recently experienced the "Golden Age of Agriculture" which was followed by the war years when they had the responsibility of providing food for Europe. After the war European farmers still struggled to recover and were no competition. American war veterans returned to the farm to build new homes, cribs, smokehouses, and barns. In 1920 the number of American farms and people living on farms reached their peak.

The agricultural prosperity soon turned to despair. European farmers started to recover and compete with the Americans. In the 1920s American farm prices fell, while the industrial sector prospered; the cost of goods that farmers had to purchase increased; farm incomes fell; and barn building slowed. In 1929, U.S. agriculture was dealt a knockout blow as the entire country fell into the deepest depression in its history. Net farm incomes averaged $450 per year (farm incomes had been almost three times higher 10 years earlier), thousands went broke, and many lost their farms. The South was hit the hardest.

This Farm Security Administration photograph of 1937 shows the disastrous effect of wind erosion in Oklahoma. (Source: U.S. Department of Agriculture)

hungry and many had no shoes. A time of despair!

Government Programs

In the 1930s, President Franklin Roosevelt set up New Deal programs designed to help farmers. Henry Wallace, Roosevelt's Secretary of Agriculture, devised the Agricultural Adjustment Act which paid farmers to limit production. New Deal programs like the Rural Electrical Administration (REA) helped provide electricity to rural areas. The availability of electricity had a tremendous impact on barn design. Both private and public utilities hired agricultural engineers to educate farmers about electricity. Barns now began to be equipped with electrically-powered lights, mechanical ventilators, barn cleaners, and water systems. These things improved labor efficiency, herd health, and milk quality. While there were slow recoveries in the 1930s, the problems of surplus agricultural production and low farm prices didn't end until World War II.

The few barns built during the Depression included small general purpose barns and feeder barns in the Midwest. Most barns were made of sawn lumber, but a few farmers in Appalachia and higher elevations of the West continued to build log barns until World War II. Sears Roebuck & Company catalogues in the 1920s and 1930s carried advertisements for precut lumber barns.

Returning to the role of early immigrants to America, one recalls their important role in barn designs. In the 1900s immigrants had little influence on barn design. They now came to a country with its own developed tradition of design. After 1920 immigration rates drastically slowed because of anti-European feelings after World War I and new laws restricting immigration. Most immigrants of the 1900s settled in urban areas.

Yet another blow came to farming in 1934. A great drought hit five states in the Great Plains — western Kansas, southeastern Colorado, northeastern New Mexico, the Oklahoma panhandle, and west Texas. Farmers hadn't practiced good soil conservation and the area turned into a dust bowl. This was a period of abandoned homes and farm buildings covered with dust. The people were

BARNS ACROSS AMERICA

A little known immigrant group that did move into rural areas in the early 1900s were Jewish. They first came to America to escape persecution by the Czars and later as refugees from Hitler. Many Jewish immigrants first settled in New York City, but later wanted to escape the sweat shops and the city life. In 1936 the Jewish Agricultural Society helped Jews relocate to farms in New Jersey, Connecticut, and the New York Catskills. They raised chickens, eggs, vegetables, and dairy products. While the Jews had ancient traditions in agriculture, in eastern Europe they weren't allowed to own land and came to America without much farming experience. They learned farming and barn building from their neighbors or through the Agricultural Society.

World War II and After

During World War II, U.S. agriculture boomed and so did the factories that made tanks, airplanes, and munitions. After the war however, America's agriculture was greatly overshadowed by its industrial power; the United States emerged from the war as the greatest industrial power in the world. The country was still one of the greatest agricultural producers, but only a small percentage of its labor force was employed in agriculture. In 1776, agriculture was the major industry; now manufacturing, construction, retailing, finance, and health services dwarfed agriculture.

The agriculture boom continued for several years after the war. But farm incomes dropped in the 1960s, then surged again in 1970 when demands for U.S. products overseas increased dramatically. This roller-coaster rise and fall of farm incomes was typical of the years beginning in 1920 and lasting through the century. When incomes rose, farmers invested in machinery, fertilizer, pesticides, and buildings. Tractors with rubber tires grew rapidly after 1950. Cotton pickers and self-propelled

Sears, Roebuck & Company sold precut lumber barns through their catalogs.
(Source: Courtesy of Sears, Roebuck & Company)

combines, and forage harvesters were introduced after 1960, and their use grew as farm incomes increased in the 1970s. This growth in mechanization was a factor leading to a more commercialized and specialized agriculture where the family farm/traditional barn was becoming a relic of the past.

Regional cropping systems were changing. The South diversified into soybeans, beef cattle, hay, and poultry. In New England farmers produced poultry, dairy products, hay and fruits and vegetables. The Midwest continued to produce corn, beef, and pork. In the central Great Plains, where they suffered from the dust bowls, farmers turned to irrigation and better conservation methods. They diversified from their intense production of wheat and grain sorghum into corn, sugar beets, soybeans, and alfalfa. In the far West, farmers depended on irrigation and immigrant workers to work the fields. Many of the farms, operated by large corporations, continued to produce basic commodities such as cotton and rice, plus fruits and vegetables.

While some regions were diversifying, individual farmers and subregions were specializing. No longer did the farmer build one general purpose building to shelter cows, horses, and beef cattle, store all their feed, and repair machinery and equipment. Textbooks and pamphlets on farm buildings published after the middle of the 20th century no longer showed plans for general purpose barns. Barns were now specifically designed for either dairy, beef cattle, swine, or poultry. Grain and feed was stored in separate large steel bins and tractors and machinery were maintained and repaired in a separate shop.

Fewer barns were being designed and built by farmers or local carpenters, more were designed by engineers and architects, and were built by contractors. Farm buildings were getting too big and too complex to be built by local carpenters; professionals knowledgeable in farm buildings were emerging. In the early part of the century, agricultural engineering was established as a profession. The Cooperative Farm Building Exchange, established in 1929 by USDA, and the Land Grant Colleges provided plans and specifications for farm buildings. Originally USDA classified their plan designers as "Barn Architects." Today, multidisciplinary professionals at the USDA, universities, and industry work on barn designs. Agricultural, biological, structural, mechanical, electrical, and other engineers as well as architects plan and design barns and other farm buildings. New types of construction also required the services and equipment of skilled contractors.

Barns were being built of new materials after World War II. Most lumber was graded and some was pressure treated to prevent rotting and insect damage. Plywood became a popular material for farm buildings. Careful attention was given to wood fasteners such as nails, bolts, plates, and gluing. Poured reinforced concrete, concrete masonry units, and precast concrete walls and floors became popular. Gypsum board, particle board, vinyl, plastic curtains reinforced with fiberglass, steel, and aluminum were used for siding. Instead of wood rafters, most newer buildings had framing of steel or wood trusses or steel girders. The late 20th-century barn designs placed more emphasis on environmental control and materials handling.

Entering the 21st Century

Farm income slowed again in 1980 when agricultural exports dropped because of a worldwide recession and the appreciation of the U.S. dollar relative to currencies of other agricultural export countries. As in other times when incomes dropped, the smaller and less efficient farmers were forced out. The high cost of land and equipment prevented young people from entering farming. Those farmers that remained had to be more prudent about their expenditures for seed, feed, fertilizers, and pesticides. They invested in large, expensive, high-tech buildings and equipment that used farm inputs more efficiently. In the 1990s, farmers started using prescription farming, a method in which they made detailed measurements of small field sections checking soil conditions and amounts harvested. People called it farming by the foot. Tractors, combines, and chemical spreaders were equipped with sophisticated sensors, navigation equipment that operated with signals from satellites and precision controls so they could do their job more efficiently. Biotechnology began to impact the industry. For example, biologically engineered hormones were approved for use in dairy cows to increase milk production.

The consequences of the economic squeeze and the emergence of expensive high-tech methods and equipment have contributed to the trend of fewer farmers and larger farms. With the 21st century there will be about 2 million farms, but only 5 percent of them produce half the nation's farm production.

Another trend in farming is the reversal of the East to West movement. Since the beginning of the nation, farmers looked westward for land and opportunities. California and the West Coast were often the goals. As we move into the 21st century, the West Coast has

These swine houses and feeding tanks are located near Milford, Utah.

been filling with people, houses, and swimming pools. Farmers and ranchers find that high concentrations of farm animals in one location creates odor and waste disposal problems, conflicting with urban development. Urbanized areas in California and the West Coast have developed stricter environmental laws regulating agriculture. Land prices encouraged the development of residential subdivisions. Consequently, there has been some eastern backtracking — farmers, for example, from California moving their operations eastward to more open areas in Arizona, New Mexico, Utah, and Nebraska.

The Modern Barn/Food Factory

As farmers move away from urban centers and try to adhere to government regulations, they are consolidating their operations into large food factories that produce milk, eggs, and meat. Our efficient national transportation system contributes to the consolidation since feed and farm commodities can be shipped longer distances. Production facilities no longer have to be near sources of feed or near consumer markets.

An example of the new food factory is a modern large swine plant in Beaver County in southwestern Utah operated by a consolidation of four companies. Earlier in the 20th century hogs were mostly produced in feed lots or pastures with small open shelters. The Beaver County facility, like most modern ones, uses total confinement. It has 2,400 sows with separate buildings for gestation, farrowing, and finishing. The buildings are large single-story structures with concrete slatted floors, steel framing, and curtain walls of fiberglass reinforced plastic sheets. And the environment is carefully controlled with fans and spray coolers. Feed is delivered automatically and the waste is removed to large lagoons.

A modern egg production factory operates about 40 miles west of Des Moines, Iowa, where hens are kept in five or six huge precast concrete environmentally-controlled poultry houses. Each house is two stories high, has cages stacked four high on each floor, and holds 120,000 birds. Automatic conveyors deliver the feed to each

This is the interior of a swine house situated near Milford, Utah.

A feed truck brings lunch to the cows at this dairy barn built in 1994 near Beatice, Nebraska.

dairy processing plants make the region a good location for dairy. In more severe climates cows are still sheltered in enclosed structures. (For more information on dairy barns, see page 75.)

These food factories have advantages and disadvantages. The major advantage is they help feed a hungry world more efficiently. The disadvantages are the mass environmental problems of waste and odors concentrated at a few locations. Animal rights advocates believe that closely confining animals in these high-density factories "is cruel and unusual punishment." (Caras, p. 119)

Several types of horse farms like general purpose, boarding riding horses, training, boarding mares and raising foals, and keeping stallions as studs require a variety of barns. Racehorse farms, for example, have created some elegant housing for horses complete with amenities such as swimming pools and air-conditioned stalls — like a five-star hotel.

cage and remove the eggs and manure. Additional buildings are used for raising the pullets, processing the feed, processing the eggs, and composting manure. The manure is bagged and shipped to urban areas for sale to gardeners.

At the T&K Red River Dairy, about 40 miles south of Phoenix, operators milk 4,000 cows. The cows are kept in large lots with shelters that shade them from the hot Arizona sun. The steel frame shelters have open sides and a roof and are equipped with fans and misters to cool the animals. Out of state producers ship hay and feed to the Southwestern dairies where it is stored in open sided sheds. Silage is kept in plastic silage bags that are 6 feet in diameter and over 100 feet long and that look like giant sausages. The feed mix is delivered to feeding troughs by large trucks. The milking is done in a milking parlor that can handle as many as 100 cows at a time. Operators are building other large dairies in the Southwest — in California, New Mexico, and the Texas panhandle. The mild dry climate, growing population, expansion of new cheese and other

Several champion studs, like Seattle Slew, live the good life in this stable at Three Chimneys Farm near Lexington, Kentucky.

This restored barn is located on the Clyde Wheeler ranch near Laverne, Oklahoma. It received a Barn Again award in 1992.

Endangered Species:
An Attempt to Save the Old Barns

Large modern food factories are replacing the old general purpose barn. Old barns, made of fine hand-hewn timbers, are on their way to extinction because of fires, termites, modern building technology, and urban development. Adobe barns of the Southwest are melting back into the earth and the precision masonry brick work of old German bank barns is crumbling. The traditional barn is a vanishing landmark.

Fortunately, there are several programs to help save and restore existing barns or to build replicas of old barns — to help save this national heritage. The National Registry of Historic Places of the National Park Service works through state and local Historic Preservation Offices to encourage the preservation of barns and other

historic buildings through tax and other incentives. The Barn Again Program, sponsored by the National Trust for Historic Preservation and *Successful Farming* magazine, encourages preservation of historic farm buildings through promotion, prizes, and publicity. For example, the Clyde Wheeler family of Laverne, Oklahoma, received a Barn Again award in 1992 for their work in restoring two old barns with leaking roofs and missing siding. They replaced much of the framing, added metal roofs, and gave the barns several coats of red paint and trimmed them in white. Perry and Gloria Clay, near Laramie, Wyoming, also won Barn Again recognition for rebuilding the foundation and reroofing their log horse barn.

Many old barns are commercially remodeled and restored into restaurants, warehouses, offices, and homes. One round barn in Champaign, Illinois, was made into a restaurant. Some farmers have converted their old barns to house specialty livestock such as llamas, ostriches, or buffaloes, or to board horses.

Living museums restore, remodel, and sometimes replicate barns and recreate historic farm communities for the public. The Museum of American Frontier Culture, in Staunton, Virginia, the Landis Valley Museum near Lancaster, Pennsylvania, and the Pioneer Arizona Living History Museum in Phoenix are good examples.

This restored log horse barn on the Perry and Gloria Clay ranch near Laramie, Wyoming, won a Barn Again award.

This round barn was originally built in 1914 for the University of Illinois College of Agriculture near Champaign. It has gone through several renovations and today it is a modern restaurant and banquet center.

The Pioneer Arizona Living History Museum near Phoenix has many examples of early Western ranch buildings such as this log barn.

Old barns, with their quaintness and charm, give us insight to a tougher but simpler life. Let us keep this monument to our rural heritage by preserving the best examples. Also, we can appreciate the new modern efficient, professionally-designed, and sanitary food factories that are contributing to feeding the world and are amazing agricultural engineering wonders.

This adobe corral holding a unique breed of horses that are descendants of the original stock brought by the Conquistadors, can also be visited at the Pioneer Arizona Living History Museum.

BARNS ACROSS AMERICA

LIFE DOWN AT THE BARN

Periodically neighbors or strangers brought massive, noisy machines to the barn to thresh grain or bale hay. Farmers held rowdy dances, solemn worship services, intense political rallies, and dramatic plays in their barns. Much of the time, however, the barn was the scene of cows and horses just shuffling about, interrupted by morning and evening chores. On the other hand, in the course of history, barns have been the site of grand and sweeping events.

This chapter focuses on barn life during the late 19th and early 20th centuries, when electricity and the internal combustion engine hadn't yet fully replaced horses as the main power source. Clever manual and horse-powered machines did many farm chores. In the Appalachia and the Rocky Mountain regions, farm jobs were performed as they had been for hundreds of years earlier; however, farmers in the mid- and far west were using new steam, electric, and gasoline powered machinery.

A COMMUNITY OF MAN AND BEAST

By the late 19th century, barns were used mostly for housing animals rather than for threshing grain as they had a century earlier. They sheltered combinations of cows, calves, horses, mules, pigs, sheep, goats, chickens, and beef cattle. More typically, the barn housed only dairy cattle and either horses or mules; other animals were kept in separate structures like chickens in a chicken coop and pigs in a pig pen.

The barn not only housed domestic farm animals, but was also a home for dogs and cats that freely moved in and out. Undomestic animals — the barn owl, the barn swallow, wrens, various reptiles, and rodents — also invaded the barn from time to time.

The stooped shoulders and downcast head reflect years of hard barn work. (Source: National Archives)

Each type of barn animal had its own personality. One writer described the farm animals as follows:

Cows — bland and mysterious
Horses — stoical
Calves — fun
Chickens — willful
Dogs — warm
Cats — aloof
Sheep — dim-witted and loveable
Pigs — greedy, dirty, smelly, and noisy (Jager, p. 112)

If there were no pigs, the barn possessed a quiet, serene, and harmonious atmosphere. Cows chewed their cud and shuffled about. Horses snorted. Another writer, reminiscing on his farm childhood, said he enjoyed sitting and milking with his head sunk into the flank of a cow and losing himself in thought (Hoy, p. 101).

The barn was a special community of man and beast. Some vague

Look who joined in this Oklahoma family portrait — the horses, of course.
(Source: Museum of the Cherokee Strip, Oklahoma Historical Society)

affinity existed between the farmer and his livestock. Farm families were fond of their animals and the animals seemed to reciprocate. Often cows and horses were given names and were treated as members of the family. However some giant farm animals weighing over a ton, such as a dairy bull, could be pretty scary beasts and working with them could be dangerous, especially for the children. But for the most part farm animals are alive, warm, responsive, and don't have negative human attributes of being critical or judgmental. This affinity between man and beast is also demonstrated in some modern nursing and convalescent homes which allow dogs and cats to roam the corridors and have rabbits, ducks, sheep, and goats in the patio yard to amuse and comfort their patients. Ann Richards, former governor of Texas, told a story about a farmer who lived alone in a most remote area of West Texas. She couldn't kill her chickens to eat because they provided her with so much company.

Of course, farm animals serve as much more than just companions. Roger Caras, in his book *A Perfect Harmony*, wrote about the important role domestic animals have played in the life of humans. They have provided milk, hides, meat, transportation, power for the plow, and companionship. He claims our civilization wouldn't have "advanced much beyond the Stone Age . . ." without domestic animals.

One of the great paradoxes is the conflict between the farm animals' role as pets and as a meat source. This conflict is dramatically told in the children's book *Charlotte's Web*. The little farm girl, Fern, loves and nurtures her pet pig, Wilbur, but her father plans to turn him into "smoked bacon and ham" at Christmastime. In the end the pig is saved. For less conflicting plots, one must read stories about orphan lambs that are bottle fed and grow to run around the farm like pet puppies.

This little girl struggled with the water pump on a Maine farm. (Source: Library of Congress)

There were many other chores in caring for animals in and around the barn that usually didn't have to be done daily but were frequent barn duties. These chores included grooming and shoeing horses and mules, nursing sick livestock, breeding selected animals, and sometimes delivering offspring.

Caring for the animals and other barn chores could be dull, hard, monotonous, routine, dirty, sweaty, smelly, frustrating, exhausting, mundane, and endless. So why has so much romanticism come out of the dust, dirt, and monotony? For one thing farmers took interest and pride in their work. They handled the routine and monotony with wit and humor. Also, nostalgia — "the past without the pain" — selectively blots out the sweat and tears. City boys and girls who remember visiting grandpa's farm particularly have fond memories of the barn and life with the animals. It was like having their own zoo. They got to help with and experience farm animals, but could quit when they became bored or tired.

THE DAILY CHORES
Caring for Animals

While beef cattle often roamed the range on their own, many farm animals such as dairy cattle, breeding stock, and horses, were housed in the barn. These animals had to be fed, watered, and cleaned up after, day-in and day-out, Sundays and holidays, early in the morning until late at night. Farm boys and girls began doing chores when they were five or six years old — gathering eggs, feeding calves and chickens, and throwing hay down from the loft. The family had to do chores during the hot, dry, dusty summer and in the bitter cold, damp winter. In the winter the children got up and did chores in the dark before school, carrying a lantern with a yellow eerie flame to the barn. By the time they were seven years old, the children were carrying water, feeding livestock, providing bedding, and cleaning stables. They drew water from the well or pumped from a hand pump. In the winter it was a particularly unpleasant chore, a careless touch of a wet hand to a cold pump could freeze the skin to the handle.

Children started working very young with simple chores like feeding the poultry. (Source: National Archives)

The city alternative to chores and farm living wasn't so great at the turn of century. Workers often had to a take long street car ride to work in a dreary factory or office. This scene depicts a Chicago street in 1903.
(Source: U.S. Department of Agriculture)

Farms were a family run business; family members shared in the many rewards and the chores — they ate, worked, and played together; only the very young and very old were exempt from barn chores. Consider the primary alternative job market, the big city. While city life sometimes offered exciting opportunities, city families often spent their lives in crowded dirty tenement houses and streets. At the turn of the century, many city workers left their families early in the morning carrying a lunch pail with a cold sandwich, took a long, shaky, noisy trolley ride to a dreary office or factory to do boring routine work for some obscure bureaucracy or corporation, and returned home much later on the same shaky, noisy trolley.

There were other positive attributes about the barn. At times it could be cozy — even a drafty barn provided protection from the winters bitter cold or shade from a hot blazing summer sun.

Many thought farming had a special spiritual quality; farmers worked with God's creation to grow food, and to feed the world. George Washington, Thomas Jefferson, and other writers and politicians have extolled the farmer's work with words like: noble, sacred, virtuous, and honorable. "Agriculture is the most healthful, the most useful, and the most noble employment of Man," said George Washington. A farm boy cleaning out a horse stable on a hot

AMERICAN AGRICULTURIST.

Agriculture is the most healthful, the most useful, and the most noble employment of Man.— *Washington.*

VOL. II.   NEW YORK, OCTOBER 16, 1843.   NO. VIII.

The masthead of a popular 19th-century magazine, "American Agriculturalist," idealized farm life. (Source: "American Agriculturalist")

summer day may have had a hard time appreciating the nobility of the task. Did these words about nobility and virtue really apply to dirty barn chores or were they really limited to sowing, tilling, and reaping the harvest?

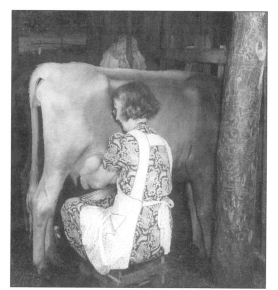

Milking

Milking was a chore often reserved for women and children on small farms. Boys and girls frequently started milking at 10 years of age if their hands were strong enough for this strenuous job. They

Hand milking, a tedious job, had to be done twice a day, every day. (Source: National Archives)

Milking machines may have increased production, but this 12-year-old Pennsylvania boy worked hard to pour milk through a strainer into a 10-gallon can. (Source: National Archives)

Sanitation was always a problem in milking and the barns had to constantly be cleaned. Particularly in the summer, bacteria could grow quickly and the workers had to clean up the waste daily and provide new straw for bedding. The walls and the floor of the milk house had to be scrubbed and the cans, vats, dippers, and brushes carefully washed daily.

Fighting Clutter

Storing, maintaining, and repairing tools and equipment were other routine jobs which took place in the barn. Farmers waged a continuous war, which many of them lost, against the clutter of tools, equipment, and farm supplies. Larger farms had sheds and a shop for farm machinery and equipment. But on most small general purpose farms, they stored equipment, tools, and supplies in the barn, in the field, or under a raised corn crib.

In the barn things were kept in spare stalls; under roof eaves; in basements; or under the cantilevered forebays. Lighter items were stored in the hay mow; heavy or large equipment such as the farm

tramped to the barn, often in the dark and through the snow, to milk the cows at 5 A.M. Milking had to be done again after school about 12 hours later. The milker would sit on the stool on the right side of the cow, first washing the udder and then milking into a bucket. It was hard and tedious and sometimes the cow didn't cooperate and would swish her tail full of cockleburs and manure into the milker's face. When the farm got electricity, milking machines eased the pain of hand milking, but one was expected to milk more cows.

The children had to help with cleanup on this Rhode Island farm. (Source: U.S. Department of Agriculture)

The barn is often used for storing and repairing machinery. (Source: National Archives)

This barn outside of Laramie, Wyoming, stored horse collars and other cowboy gear. (Source: National Archives)

wagon, and later tractors and pick-up trucks, were kept in the barn alley or on the threshing floor along with plows, cultivators, and hay rakes. Equipment for processing grain and other crops was also stored in the barn. They included threshing flails, threshing machines, winnowing machines, hammer mills for crushing corn into feed, turnip choppers, root slicers, straw cutters, potato washers, and corn shellers.

Other equipment stored in the barn included devices for bagging and handling the grain or other commodities: sack holders, mechanical rope twisters, grain samplers, bushel measures, scales or weighing machines, wheel barrows, scoops, pitch forks, and shovels. There were ropes, blocks and tackle, and wire that hung from the barn walls or beams. Barns with cows had the familiar metal 40-quart milk cans setting about. There were collars, bridles, harnesses, shoes, and blankets for the horses or mules. Today one can still find a few relics of the past in old barns. One finds these old items among artifacts of today's world: plastic lawn chairs, motor bikes, truck tires, steel filing cabinets, skis, ping pong tables, and abandoned computer terminals.

ROUTINE IS BROKEN

Periodically, exciting events broke the daily routine at the barn. These activities included hay baling, grain threshing, business wheeling and dealing, dancing, and even praying. New life was generated in the barn with the birth of calves, lambs, and foals.

Grain Threshing

Threshing small grain such as wheat and barley was an important barn event. The word "barn" comes from an old English word meaning barley. Farmers cut the grain in the summer and bundled sheaves together into shocks which were allowed to dry in the field. The shocks were brought to the barn in a wagon, forked off, and stored in grain mows. The grain could be threshed later; often as a winter chore. In early threshing operations, the grain was separated from the straw by hand flailing. Workers laid out the sheaves of grain in rows on the barn floor. Threshing floors had to be wide enough to hold the horses and the wagon. After the sheaves were laid out, the farmer used a flail, a paddle-like tool that was free to rotate on a long handle. The workers would simply raise the handle and beat the grain with the flail. Workers kept up a steady rhythm pounding the grain.

After the grain was threshed it was winnowed. This chore was done on a day when the wind was moderately strong. The wide barn doors were opened to let in the breeze. Baskets filled with the threshed grain were tossed up and the wind blew away the chaff.

In another threshing method, called treading out, the grain was threshed by animals hooves. A child would often ride one horse and lead another in a circle treading over the sheaves of grain. Treading out was sometimes done in a barnyard or field where the workers laid out the sheaves on a canvas tarp. The flailing method recovered more grain and was subject to fewer impurities than treading out. With treading out, the farmers tried to keep the manure off the grain by using a leather bag on a stick to catch the manure. George Washington's special treading out barn can be found in the section on round barns on page 69.

In the mid- and late 1800s many mechanical threshing machines were developed which used mechanical flails operated by horse power. Two to four horses attached to a sweep walked in a circle and through a system of gears, shafts, or belts, the horses powered the threshing machine. Another horse-powered

Steam-powered machines threshed beans in southern California. (Source: California Historical Society, Neg. 2270)

machine was the stationary treadmill popular in Pennsylvania. One or two horses walked on an inclined continuous belt much like the treadmills used today by people for exercise. The mechanical threshers and other processing machines of the day offered Rube Goldberg-like devices with levers, belts, gears, pulleys, ropes, bins, sieves, and trays. They sputtered, clanked, and shook, usually operating in a cloud of dust. But out of the end of the machine came the threshed grain, shelled corn, or chopped straw. These machines were not too delicate in handling the commodity — a far cry from modern machines that can handle vine-ripened tomatoes. They weren't safe either; the moving belts, gears, teeth, and knives without guards were an ever-present danger. Horse-powered threshing was replaced in the major grain producing regions in the late 1800s with steam powered machines and later by the self-propelled combines powered by internal combustion engines.

Two horses on the treadmill provided the power for a grain thresher. These machines were popular in the 1800s in Pennsylvania. (Source: Library of Congress)

By the late 19th and early 20th centuries most threshing had already been converted from a barn to a field operation. Grain production had moved into the arid western plains where threshing was done with field machinery. However, flailing and treading out, the methods used in early colonial times, were still used until World War II on a few small general purpose farms in the Northeast, the Appalachian region, and the South.

Whatever method was used — failing, treading out, or mechanical threshing — grain threshing was dusty, dirty, itchy, lung-irritating work. Today, the only hand flailing or horse treading-out is done as demonstrations at county fairs and craft shows or by a few conservative Amish farmers.

Haying

As farmers kept more livestock, they fed more hay, and hay storage became a major part of the barn's operation. Farmers liked to feed hay — the stems and leaves of grasses and legumes — because it provided the most nutrients for the lowest cost. With more emphasis on feeding hay, barn designs changed. They were built with bigger hay lofts. One way this was accomplished was by building barns with high gothic or gambrel roofs. For access to the loft, big doors protected by large hoods were built at the ends of the hay loft. To handle the giant loads of hay, tracks were erected at the top of the loft along the roof ridge. The tracks carried a mechanical trolley with a grapple hay fork or other device for picking up and moving hay loads.

Hay had to be cut at just the right time — too early and it was full of sap and hard to cure, too late and it was too woody and lacked nutrients. Because the cut hay had to cure in the field for one to three sunny days, the selection of the cutting day was critical. Even up until World War II, many farmers still cut hay with a horse-drawn sickle bar mower and raked it with a horse-drawn dump rake. This rake had a row of large curved tongs or teeth that combed the hay field. When it was full, the farmer pulled a rope lifting the teeth and dumping the cut hay. The cut hay was loaded on a horse-drawn wagon with pitchforks and then it was taken to be stored loose in the barn.

During the 1930s the side-delivery rake, the hay loader, the tractor, and the truck were introduced. Barns needed bigger hay mows to hold the larger hay crops. The side-delivery rake had a rotating reel set at a 45-degree angle with the line of travel. The reels delivered the hay to the end of the rake and formed a continuous windrow.

In 1875, hay was cut by a hand scythe and a horse-drawn mower.
(Source: Library of Congress)

In the 1930s farmers started loading hay on their wagons and trucks with a mechanical hay loader.

Farmers didn't have to stop or slow to dump the hay. The windrow formed by these rakes worked well with the new hay loaders pulled behind a wagon or truck. The loader was a wide conveyor belt on wheels with one end of the conveyor belt near the ground. A rotating cylinder gathered the hay and fed it onto the bottom of the conveyor. The conveyor belt slanted forward at a steep angle with its top high above the truck or wagon. The belt carried the hay up and dropped it into the wagon. Horses or mules had to really strain when pulling the hay wagon plus the loader.

The wagon was hauled to the barn where it was maneuvered under the barn hood (the covering over the opening to the hay mow at the end of the barn) or into the center of the drive floor. A large harpoon or grapple fork was plunged into the hay, picking up a substantial portion of the wagon load with one thrust. Some farmers used slings made of rope and wood staves. They were placed on the wagon in the field and the hay loaded on top of them. The fork or sling load of hay was lifted up to the barn peak with a rope mechanism usually pulled by a team of horses but sometimes by several men, a tractor, or a small motor. When the load reached the

top of the barn, the fork or sling locked into a carrier that operated on a track running just under the roof ridge and along the length of the barn. The carrier was pulled along the track to its correct position and then a trip rope was pulled releasing the load. Workers could also give the load a big swing before releasing it to get it into the barn corners. Workers using pitch forks spread out the hay to fill the barn evenly.

Loading the barn on a sweltering July day was hot work. One writer said when you get out of the top of a barn you "could wring water out of your stocking" (Hastings and Hastings, p. 46). Workers carried a half gallon or gallon crock water jug with a corn cob stopper. Even a lukewarm drink from the jug was a satisfying relief when working in a hot hay loft.

From the 1920s to early 1940s the storage method changed as hay baling started to replace loose hay. The first balers were stationary units driven by horse power and set up in the barnyard or the hay field. One or two horses or

A hay fork on a block and tackle was used to load the loose hay into the barn.

During the 1930s farmers started using horse or tractor powered stationary hay balers in the field or in the barnyard. (Source: Library of Congress)

mules, harnessed to a sweep, walked in a circle. The sweep drove a piston that forced and compacted the loose hay into a chamber. Later balers were belt driven by a steam engine or tractor. A bale about the size of a small trunk weighing from 50 to 150 pounds was ejected from the baler. These are called square bales; while they are more or less square if you look at the end, they are rectangular if viewed from the side or top. Workers hoisted the bales into the hay loft with ropes, pulleys, and a carrier similar to the ones they used for loose hay. However, instead of lifting with a fork or sling, they used hooks or tongs. Later, mechanical conveyors replaced the block and tackle. Much more baled hay could be stored in a loft than loose hay. Some old barn floors couldn't carry the heavy load of baled hay. Farmers built new larger and stronger barns or tried to reinforce the existing hay loft floors to accommodate the heavier loads.

Today, tractors pull mobile balers through the field that pick up and bale the hay. Bales have become larger and heavier. Some larger square bales weigh 1,000 pounds or more and are about the size of a pick-up truck bed. They look like a giant shredded wheat biscuit. Round balers produce cylindrical bales which are about 6 feet in diameter and 5 feet thick and weigh about 2,000 pounds. The larger bales are often covered with plastic or canvas and left in the field. The so-called round bales shed rain water and frequently are not even covered. Field hay storage is

The bales were loaded into the loft with tongs or hooks and a block and tackle. (Source: Earl Thollander)

Catching the hay bales in the loft coming off the conveyor was one of hottest jobs in the barn. (Source: U.S. Department of Agriculture)

reducing the use of the old-fashioned hay mow, particularly in the arid West. During the first part of this century in the West, loose hay often was not stored in a barn but in giant stacks in the field. Horse powered stackers and derricks piled hay in giant conical-shaped stacks. Today the hay stacks are gone, but in southern Utah, Nevada, Arizona, and New Mexico, one may see sun-bleached bales of hay, which look more like straw to an Easterner, stacked in open corrals and fields or covered by sheds without sides.

Large round bales were stored in the barn lot at the 24,000-acre Gross family ranch at Pine Bluff, Wyoming.

Peddlers and Other Visitors

Buyers and sellers periodically pulled up to the barn in their wagon or truck to "wheel and deal" with the farmer. Trading was closely associated with the barn. Traders bought and sold barn products: cattle, calves, horses, and hay. They bought and sold machinery and hardware that was stored in, or was a part of, the barn. Even when dealing with non-barn items, the negotiations often took place in and around barns while the farmer was doing his or her chores. Buyers came at regularly scheduled times to buy eggs, milk, or cream. The more mysterious and exciting buyers were the strangers who showed up unexpectedly seeking to buy poultry, cattle, hay, timber, mineral and oil rights, or whatever else one could find on a farm.

Peddlers came with their wagon loads of not only small farm equipment, but also household items: tinware, kitchen utensils, notions, Dr. Kilmer's kidney cures, Watkins' veterinary liniments, trinkets, Bibles, and other religious books describing last day apocalyptic events. Sometimes the farmer paid for purchases by providing dinner and overnight lodging.

Not all buyers and sellers were honest. Some tried to swindle the farmer. Rainmakers operated on the plains into the 20th century. In the 1920s and 1930s, Agricultural Extension Agents encouraged farmers to cull their poultry flocks and herds to get rid of the disease-infested birds or animals or the poor producers. Enterprising and unscrupulous people took advantage of this well-meaning educational program. Men showed up at the farm representing themselves as from the "State Poultry Association" or the "Livestock Improvement Association" coming to help them cull or improve their poultry or livestock. They might find all the farmers flock infested with a rare disease, but fortunately they just happened to have an effective but expensive medicine in their wagon that could cure the problem. Or they may have tried to sell the farmer an ordinary sow representing it as a special high-producing breed for more than four times its worth. Today the old peddlers are for the most part gone. Now buyers can sit in the comfort of their home office and bid on cattle through TV auctions.

It was an intense day whenever peddlers and buyers arrived at the barn, as seen in this picture called "The Poultry Buyer." (Source: "American Agriculturalist," Dec. 1875)

These boys played in a spreader stored in the barn. (Source: U.S. Department of Agriculture)

Playing, Partying, and Romancing

It wasn't all work and business around the barn. A hay loft of either baled or loose hay was a delightful place for children to climb, jump, and hide. The hay smelled sweet and felt comfortable. Boys and girls climbed the ladder in the loft reaching the highest tie beam, then jumped or swung on a rope, and fell into the loose hay. If hay bales weren't too heavy and the children were strong, they built tunnels or castles with them.

The barn was a hide-out for naughty and the near naughty adventures. Young boys searched for the white flower of a weed in the hay called rabbit tobacco. They rolled the flower in a newspaper to make a crude cigarette or stuffed it into a homemade corn cob pipe. It was good for a smoke, and was a lot more benign than the real thing. As for the real thing, many a boy, and a few girls, sneaked smokes in back of the barn. Dads, uncles, and older brothers might sneak a drink of "white lightning" behind the barn, even though the women were members in good standing of the Women's Christian Temperance Association.

Young couples often found the barn was the only place they could be alone and get acquainted. One of the more famous barn romance scenes came from the early 1940s movie, *The Outlaw*. In this movie there was a famous provocative barn scene of Jane Russell and Jack Beutal "Billy the Kid" in the hay. Pictures and stories like this have helped create the amorous figure of speech associated with barns: "rolling in the hay." One real life barn romance turned out tragically. In 1904, Dolly Douthitt of Enid, Oklahoma, shot her husband when she discovered him and the hired girl romancing in the barn (Porter).

The barn was a place for work bees combined with entertainment. At corn husking bees, the neighbors gathered on the large threshing floor to husk hundreds of bushels of corn. Sitting on crude

Jane Russell and Jack Beutal "Billy the Kid" got acquainted in the hay in the movie "The Outlaw." (Source: Courtesy of the Academy of Motion Pictures Arts and Sciences)

Grandma Moses: "The Barn Dance" Copyright © 1969.
(Source: Grandma Moses Properties Co., New York)

rural people took their music and dances with them as they commingled with Northerners in the army and defense plants. Roy Rogers, Gene Autrey, and other cowboy stars spread their type of western music by way of the movies. More recently cable TV brought country line and couples dancing into the homes of America. The result is a growing popularity of country western music and dances. At the same time, suburban populations are surrounding old barns. City dwellers and suburbanites like to shed their business suits for Durango boots, blue jeans, and Stetson hats, to party and dance in a real or make-believe working barn. They wear cowboy shirts or tee-shirts with words and pictures depicting: horses, pick-up trucks, saddles, bulls, racing cars, Nashville, barns, cowboys, beer, rodeos, or guitars. Lawyers, doctors, and engineers gather Saturday night in barns with construction workers, auto mechanics, and farmers to dance the *Boot Scootin Boogie, Slappin Leather*, and the *Cowboy Waltz*. The barn's floor is slippery and dusty; walls and roof are covered with cobwebs; music is loud; and the only place to sit is on bales of hay or straw. Instead of a fiddler, the music is usually played by a disc jockey using compact discs played on high-tech electronic gear blasting out sound through several megaspeakers. The dancers "party all night till the cows come home."

More Serious Activities

The Old Order or Amish had a tradition of not worshiping in churches. In Europe they met in homes and other clandestine locations to avoid recognition and persecution that could mean jail or death. In America, they settled in Pennsylvania, Illinois, Ohio, and Indiana. In these areas they had to endure less serious harassments. Even today some Amish orders continue the tradition of meeting in homes in the winter and in barns when the climate is milder. Consequently, people call them the House Amish; one could just as well call them the Barn Amish.

benches around a huge pile of unhusked corn, they husked the corn and tossed the ears over their shoulder into bins. They turned the work into a game by giving prizes for those finding a red ear. Usually cookies, cakes, and sandwiches were served at these events.

Social events such as community sings, parties, and weddings also took place in the barn. Dances were popular, such as the square dance, except in communities of strict religious sects. At the dance there would be a caller yelling out the "do si does" and "left alamans" and the fiddler or guitar player providing the music. At Christmas, and especially Halloween, parties are still held in barns.

In the second half of the 20th century there has been a resurgence of the barn party and dance. During World War II, Southern

Men with chin beards and dressed in black collarless coats, women in bonnets, and boys and girls, sat on crude wood benches in the barn. They were separated by gender and age. The seats next to the walls or posts were taken first because they provided back support during the long service. The setting was simple and austere — no bells, icons, incense, or crosses — and the leaders sat at a long table in front of a small congregation. There might have been a bucket of drinking water and dipper, otherwise it looked like an everyday working barn. Traditionally there were two

It's not only the Amish that have barn weddings. Occasionally couples of other traditions get married in a barn. Molly and Robert Ellis of Philadelphia were married in her father's barn in Montgomery County, Maryland. (Source: Molly Ellis)

sermons, usually preached in a combination of English, high German, and Pennsylvania German, long scripture readings, and singing 16th-century hymns in slow solemn monotone chants. The service might last for several hours. In addition to regular church services, the Amish held weddings, funerals, auctions, and other community activities in barns.

Sunday and other evenings the Amish often had (and still have) singing or "frolics" in the barn to help the young people work off steam. Music was provided by a harmonica player or an old-fashioned phonograph. There might even be square or folk dancing, but no modern dancing that could seduce the young people. Recently,

Chairs were set up in the Roslyn Farm barn in Montgomery County, Maryland, for the wedding. (Source: Molly Ellis)

however, young people have been holding dances without their parent's consent where hundreds of teenagers get together in a barn for modern dancing to an electronic guitar or tape player.

There is a famous painting by John Curry, an early 20th-century painter of Midwestern farm life, called the *Baptism in Kansas*. It shows people being baptized in a large steel livestock watering tank

These Amish boys were at a cattle auction in eastern Ohio. (Source: Fred J. Wilson)

ONLY RARELY

While most barn activities were part of a daily routine or periodic events, occasionally there was the historic or rare event associated with a barn.

The Beginning and the End

Two rare events in the life of every barn were its erection and destruction — its birth and death. Often the barn was built as the result of a barn raising — where a hundred or more people would gather to pull on ropes and push with pikes or large poles to erect the heavy wood barn framing. It was an exciting and happy occasion. After the raising, the workers and families gathered around large tables for a home-cooked feast.

The end of a barn could come about because of fire, termites, developer's bulldozer, vandals, or high winds. Fire was a particularly serious problem. Lightning and spontaneous combustion of hay have been leading causes of barn fires that have not only destroyed

in a barnyard with a large western barn in the background. This is additional evidence that barns have been an important part of American religious life.

Besides religious meetings, the barn has been the scene of community gatherings. A newly organized Grange group would meet in a barn before they built their hall. They held heated discussions about the need to reform the banking system and railroad rates. The barn would also serve as a hotel for temporary hired hands or transients who would sleep in the hay mow or an empty stall. The term "barn storming" means to travel around giving plays, political speeches, and lectures. At one time, politicians, performers and snake oil salesmen, used the barn and the barnyard as their platform.

John Steurt Curry: "Baptism in Kansas" 1928.
Oil on canvas, 40 x 50 inches (101.6 x 127 centimeters)
(Source: Collection of Whitney Museum of American Art,
Gift of Gertrude Vanderbilt Whitney)
Photograph Copyright © 1997.
Whitney Museum of American Art

This Amish barn raising was in eastern Ohio. (Source: Fred J. Wilson)

This barn raising was in the National Building Museum.

the building but also the valuable contents including animals. Many barns had several lives because they were rebuilt several times or the materials moved to a new location to make a new barn.

An even sadder event than the destruction of the barn was the farm sale. The barn might still be sturdy, but the property often had to be sold because the farmer had died, was too sick or old to keep the farm, or just couldn't make the mortgage payments. The farm might be sold to be turned into a housing or industrial development. Sometimes the sale may have been a happy time when the family was moving to a bigger and better farm or to retirement in Florida. Even so, there was usually something sad about leaving the old farm.

Farm sales were usually in the early spring before new crops were planted or in the fall or early winter after the harvest. Several hundred people attended a sale and refreshments would be served. The auctioneer may start in the barn or barnyard auctioning off the cattle, mules, harness, cultivator, corn shellers, and all the rest. A good auctioneer would tell stories or jokes to keep up interest during lulls in the bidding.

The sale often meant the end of the barn. (Source: U.S. Department of Agriculture)

End of The Biggest Manhunt

One of the biggest manhunts in American history ended as the fugitive was shot in a barn. It was a tobacco barn on a farm in the tidewater region of Virginia, near the town of Port Royal, 50 miles south of Washington, D.C. The barn was part of a farmstead located among pine trees and an orchard with several outbuildings. Federal troops surrounded the barn and eventually set it on fire to drive out the fugitive. He was shot fleeing the barn. The year was 1865; the fugitive, John Wilkes Booth.

After shooting Abraham Lincoln and breaking his leg jumping off the balcony at Ford's Theater, Booth rode out of Washington on a mare. He teamed up with Davy Herold, a drugstore clerk familiar with southern Maryland, and together they continued south. After getting treatment for his broken leg and hiding out at various locations, Booth and his comrade crossed the Potomac River to Virginia. They ended up at Richard Garrett's farm, where Booth posed as John W. Boyd, a confederate soldier wounded at the siege of Richmond, and Herold posed as his cousin.

The farmer let them stay in his house but after becoming suspicious, he asked them to leave but allowed them to sleep in the barn until morning. At 2:00 AM, 26 enlisted men of the 16th New York Calvary, along with officers and federal detectives, arrived and surrounded the barn. Herold surrendered but Booth tried to negotiate with the officer in charge. Speaking in a stage voice, he alternated between using flattery: "Captain I believe you to be a brave and honest man." and threats ". . . I will fight your whole command." (Smith, p. 211) The soldiers torched the barn. They could see Booth lighted by the flames through the

John Wilkes Booth was shot as he fled the burning barn. (Source: Library of Congress)

wide ventilation cracks in the barn wall. One leader of the group compared Booth, standing in the light of the fire watched by soldiers, to a stage performance in front of footlights before an audience. Booth dropped his crutch, and holding his pistol, he ran to the door. Against orders, Sergeant Boston Corbett, a mad religious fanatic, shot Booth. He died on the veranda of the house muttering the words "useless."

A Barn Fire Destroys a City

The worst fire in the world's history started in a barn in Chicago on October 8, 1871. The O'Leary's had a barn about a mile southwest of the central business district in Chicago on a small lot near their house where they kept five milk cows. Catherine O'Leary milked the cows each day and delivered milk to neighbors. Three tons of timothy hay was delivered and stored in the barn's loft early the day of the fire. About 9 P.M., flames were first discovered shooting out of the barn. The strong dry winds fanned the flames north and eastward for 4 miles devouring over 17,000 buildings, mostly of wood. Homes, factories, department stores, offices, grain elevators, hotels, mansions, sheds, and other barns were consumed. The fire killed 250 people and left 90,000 homeless. According to legend the fire started while Mrs. O'Leary was milking a cow that knocked over a kerosene lantern. The actual cause of the blaze is unknown; Mrs. O'Leary was in bed at the time the fire started. But Mrs. O'Leary and her cow became the villains and were blamed for the great fire. The O'Learys eventually fled the area.

Mrs. O'Leary and her cow became the villains of the great Chicago fire and were the subjects of nasty stories and cartoons. Mrs. O'Leary was home in bed when the fire started. (Source: Library of Congress)

Entered according to Act of Congress, in the year 1871, by H. M. Kinsley, in the Office of the Librarian of Congress at Washington.   Painted by L. V. H. Crosby.

ORIGIN OF THE CHICAGO FIRE.

OCTOBER 9th, 1871.

Old Story Generates an American Custom

In contrast to these tragic and sad barn events, there is in the Christian tradition, the good news and glorious story of the birth of Jesus in a stable. "And she brought forth her firstborn son and wrapped him in swaddling clothes and laid him in a manger; because there was no room for them in the inn." The story includes an account of wise men presenting "unto Him gifts; gold, frankincense, and myrrh." (Luke 2:7) It happened a long time ago in a land many miles from America, but the story probably led to an American rural tradition in some communities of hiding Christmas gifts in the barn. Perhaps this tradition was reinforced by Scandinavian immigrants. In their folklore, a barn Yule elf lived in the loft at Christmas. The housewife would set out a bowl of porridge for him and he would bring presents for the children (Gardner, p. 20).

Christmas has long been associated with the farm animals. St. Francis of Assisi long ago said that the animals must be a part of Christmas rejoicing because of their role at the birth of Jesus. In some traditions, the people tell about the cows breathing their sweet breath on the baby Jesus to help keep him warm. In many Old World traditions the cows, pigs, and horses have the power of speech at midnight on Christmas eve, but it is most dangerous for humans to eavesdrop on their conversations. If they do, they will suffer serious bad luck. The barn animals often stand or kneel towards the East in honor of the Christ child. In turn, farmers pay homage to cattle at Christmas. In old English traditions, the family comes to the barn and honors the animals with a special ceremony raising their cups and toasting the animals with ale. And to help the animals rejoice, they were given extra hay and corn and sometimes even a taste of strong drink.

This idealized nativity scene with a stable, "The Adoration of the Magi," was painted about 1445 and is considered among the greatest of Florentine paintings. It was probably a collaboration of two artists, Fra Angelico and Filippo Lippi.

(Source: Samuel H. Kress Collection)
Copyright © 1997 Board of Trustees, National Gallery of Art, Washington.

DESIGNS FOR THE TIMES

Barns vary dramatically — there are square barns, English barns, Pennsylvania bank barns, Wisconsin dairy barns, single pen barns, Dutch barns, round barns, balloon frame barns, and heavy timber barns, to name a few. Barns can be classified in many different ways: by their ethnic origin, construction, pen configuration, or shape. While some barns in America carry the name of a foreign nationality, they were all built in the United States. Some bear the name of a state, but they were often built far beyond the state's borders. Many barn types are not clearly defined and some are mixes of two or three types. It's difficult to select a representative cross section of American barns. Consequently, this chapter focuses on five great diverse American barn designs: the log barn, the German or Pennsylvania bank barn, the Southwest adobe, the round barn, and the modern dairy barn.

MASTERING THE FRONTIER WITH LOGS

Log construction is a romantic symbol of the courage, strength of character, resourcefulness, and adventurous spirit of American settlers. People born in log cabins are considered honest and genuine Americans. Ironically, while the log construction has been symbolic of honesty, much fiction surrounds it. Frequently quoted examples of this fiction are the pictures of pilgrims trudging to church in the snow past their log buildings and the 1840 presidential campaign of William Henry Harrison implying he was born in a log cabin. The pilgrims didn't build with logs and Harrison was born in a Virginia mansion.

The trademark of log construction is the horizontal solid wood timber. This arrangement is in contrast to the post and beam construction where the main members are timbers set vertically and spaced about 16 feet apart. Lighter horizontal members spanned between them and this entire skeletal framing was covered with thin wood sheathing or shingles. Log and frame were the two main types of wood buildings in early America.

Years before the Europeans came to America, the Indians built log cribs using small unnotched logs. While the English brought their language and laws to America, they did not bring log construction. There were no log buildings in England, Ireland, Scotland, or Wales. Conventional log barns were first built in the 1630s by the Swede's and the Finn's who settled New Sweden in the Delaware Valley in what is now New Jersey, Delaware, and Pennsylvania. Their settlements were small and isolated; they apparently had only a moderate role in spreading log construction.

The German settlers in southeastern Pennsylvania and those from Switzerland and the present Czech Republic have mostly been credited with spreading log technology. They settled in Pennsylvania in the early 1700s and built log cabins and barns. Their methods were adapted by the English and the Scotch-Irish and log construction spread southward and westward. The English settlers didn't build many log barns in New England or in southern Tidewater. From the southeastern Pennsylvania core area, log construction spread into Maryland, Virginia, southern Ohio, and throughout the great forested areas of the East. Other ethnic groups had minor roles in spreading log construction. These included the French in the Great Lakes and

This drawing illustrates the distribution of log construction in the eastern United States. (Source: Noble, Vol. 1)

Mississippi Valley, the Russians in the Pacific Northwest, and the Spanish in the higher elevations of the Southwest. In the mid-1800s the Finns who settled in the northern woods of Minnesota, Michigan, and Wisconsin also built log barns.

Log barns continued spreading across the country. Farmers built log barns in Iowa, Illinois, and Indiana, but they were somewhat limited. Not only were there few large trees, but farmers also needed bigger barns than they could build with logs. For the most part, the farthest west that log construction went was eastern Texas where it reached in the 1820s. Further west the trees were too small and sparse, and settlers built few if any log barns in the Central Great Plains which ran from west Texas to North Dakota. Some log barns were built in the far West in the higher elevations where there were more trees.

In eastern Tennessee, eastern Kentucky, and western North Carolina, farmers were building the cantilevered Tennessee barns in the late 1800s and early 1900s. Log barns continued to be built during the Depression and even up to World War II particularly in Appalachia and in northern Utah and Idaho, in heavily wooded areas with marginal farms

Expanding the Barn with Cribs

Settlers assembled logs to form small rectangular pens or cribs about 12 feet long by 8 feet wide. The logs were joined together at the corners with notches. The crib (or pen) is the basic unit of log construction. Cribs typically had a gabled roof with the ridge running lengthwise plus a door in the gabled end, and were used for storing corn, grain, or hay or housing cows, horses, pigs, or chickens.

As farmers expanded their operations, they built larger cribs although tree size limited crib size. The larger cribs, mostly in Pennsylvania and Appalachia, were 18 feet by 18 feet, however a few cribs have been reported as large as 30 or 40 feet square. Farmers added more barn space by building several separate cribs or adding sheds to the sides of the central log crib. Animals were often kept in the shed and grain stored in the central crib.

Still another method of expanding the barn was to build two cribs about 14 feet by 14 feet, separated by another 12 or 14 feet, and then building a common roof over

The Rule barn in Sevier County in eastern Tennessee is an excellent example of a cantilever Tennessee double crib barn.
(Source: University of Tennessee)

This single log crib was built about 1891 and now stands at Heritage Farm Museum, in Plano, Texas. It was built off the ground on wood poles, but cribs were typically built directly on the ground or on stone foundations.

This small double crib, made of hand-hewn logs, was built around 1840 in Washington County in southwestern Virginia.

the two cribs. The area between the cribs sometimes had a wooden floor used for threshing grain, but may have only had a dirt floor which was used simply as a passageway for bringing in wagon loads of corn or hay. In the double pen arrangement, one pen might have been used to house animals and the other for grain storage. Above the cribs and central passageway was usually a mow for storing hay and straw.

One type of double crib barn was the Tennessee cantilevered barn in which the hay mow and roof projected beyond the cribs. These overhangs usually ran beyond the front and back of the cribs, but sometimes barns had overhanging projection on all four sides. The farmer often stored plows, tools, and machinery under these overhangs. A few of these barns are still used in southern Appalachia for curing burley tobacco. Farmers have added supports

These farmers were building a single crib potato house of unhewn logs, in Natchitochas County, Louisiana. (Source: National Archives)

This triple log crib was built around 1800 in Washington County in southwestern Virginia.

to the cantilevered overhang of many barns or added walls that enclose the area under the overhang thereby disguising the distinct feature of these unique barns

Farmers also built triple cribs that were similar to the double crib but the center passageway was closed off to form another crib. Some log barns had four cribs with two intersecting driveways. Farmers converted the four cribs into six by closing off one driveway and converting the space into two more cribs. This provided one central driveway with three cribs on either side and was called a transverse barn. In any of these barns, sheds were often added to the crib sides.

Construction

The first step in building a log barn was to clear the land of trees and shrubs. Next the builder laid a foundation of field stones. Trees — usually oak, hickory, or poplar in the East and pine, cedar, or other coniferous trees in the West — were cut for the walls. Oak and other hardwood were more durable, but pine was easier to work with.

The bark was customarily peeled off the logs, making them easier to work with and providing a smooth wall surface. Rarely did builders cut logs more than 24 feet long. Longer logs were too heavy

to handle and the taper was too great to provide a good straight piece, or working structural member. Usually logs were 12 to 18 inches in diameter and one or more sides were squared off or hewn with an axe. For crude barns, builders left the logs round, but for better barns, builders typically would hew the logs on two sides — the sides for the interior and exterior walls which left a wall thickness of 6 to 13 inches. The top and bottom of the log would be often left round. Logs for houses and for the best barns were hewn on all four sides.

To assemble the logs, the farmer obtained the help of four or five neighbors and together they could raise the barn in one day. For the most part, men did the hard labor while the women helped with lighter work. Builders started by placing a large, straight log, called the sill, on top of the foundation. If the building was to have a wood floor, the sill was notched to fit the floor joists. Logs were added on top of the sill to form the wall. As the wall got higher, the logs were slid up skids placed on top of the wall. Sometimes women used horses to pull logs up these skids with rope.

The key to log construction was building the notched corners. The better craftsmen worked as corner men — usually one man on each corner. Unnotched logs, which were longer than the wall, were

This early woodcut engraving shows a single log crib raising. (Source: "Western Miscellany," July 1848, Vol. I)

temporarily slid into place on top of the wall. Then they were cut off to the proper length. Next, the undernotch was cut; the log was rolled into place; and finally, the over-notch cut.

The Notches. Generally, the type of notch depended on the kind of wood, the builder's skills, and the ethnic background. The three common notches were the saddle, the "V," and the dovetail. The saddle notch is the oldest and simplest and was used in early pioneer cabins, barns, and outbuildings. Unskilled workers, using an axe, could easily cut one or two crude circular depressions on each end of the log. It was most common in logs that were left round and often was used with soft wood such as pine and cedar.

The V-shaped notch is a variation of the saddle notch; instead of a circular cut, the notch is V-shaped. German settlers commonly used the "V" notch in Pennsylvania, and it spread into Appalachia and the Ohio Valley. The "V" notch was used both on hewn and round logs.

The dovetail notch, normally used with the hardwood and hewn logs, is formed by cutting slants or splays in different directions at the end of the log. (One can often find dovetail construction in the joints of an antique dresser drawer.) A full dovetail has two slanted notches at the end of the log — one on the top and one on the bottom. The half dovetail has one slanted notch cut on top. Dovetail

The "V" notch was a popular notch for barn builders.

notches have superior strength and the direction of the slant forces water to drain to the exterior of the building preventing decay. The full dovetail had limited use in log construction since it was time consuming and required special skills to construct. It was used in the Delaware Valley, eastern Pennsylvania, and Appalachia. The half dovetail was more common in Virginia, West Virginia, throughout the Ohio Valley, and further west into Arkansas and Texas.

Laying Up the Walls. During construction, builders first laid up solid walls without putting in openings for windows and doors. Where openings were to be added, the logs were half notched at the top of the planned opening so the builder could insert a saw. Logs around the proposed opening were blocked with wedges. The saw was then inserted in the half notch at the top of the opening and the logs were cut. After they were cut, the builder would insert jam boards against the sides of the opening and fasten them to the logs with wooden pegs.

The tops of the log walls were capped with a hewn plate. These plates held the roof rafters that were 6-inch poles or sawn lumber. The rafters were usually covered with 1-inch sheathing and then with shingles, shakes, or clapboard. Because the logs were uneven or warped, there were 1- to 6-inch spaces between the logs making up

Dovetail notches had superior strength.
(Source: The Kentucky Heritage Council)

the wall. Warping often occurred when the logs were put up green. Sometimes these spaces were left for ventilation in hay storage or tobacco curing cribs. Usually, the spaces, called chinks, were filled with stone, small boards, clay, mud, or lime and sand mortar.

In conventional log construction, all of the building's weight rested on the corners. The exceptions were the Scandinavian log buildings located in the upper Great Lakes. They hewed logs with a convex top and a concave bottom. The concave bottom of the log fit snugly into the convex top of the log below. The logs rested on top of each other for their full length.

The French, in Wisconsin and elsewhere in the upper Great Lakes, erected another variation of the log barn called "piece-sur-piece." The logs were hewn square with a tenon in each end inserted into grooves cut into an upright corner post. It made a very weather-tight house or barn.

Logs played an important role in American barn building because farmers could use the abundant supply of local timber. The main tool needed was the axe. A settler without much skill could put up a barn quickly on the American frontier. The locking joints of log construction created a very stable structure. The expansion of saw mills and cut nails and the development of the railroad that could carry lumber to all parts of the country helped bring an end to the log barn. As farmers raised more crops, they needed bigger barns. Log construction was not adapted to the large barn. A few log barns continued to be built throughout the Depression and until World War II. But their romanticized image standing for integrity, honesty, courage, and resourcefulness finally gave way to a stigma of the crude and the old fashioned.

PLAIN PEOPLE ERECT SOPHISTICATED BARNS

One must admire the great workmanship of the massive timber frames. The barns are pleasing to the eye and are durable. It is no wonder that these barns that started in the southeast corner of Pennsylvania, spread across the country. Farmers built them for 200 years. These large timber-frame barns are called German bank barns or more accurately Swiss German bank barns. They are also called Pennsylvania barns or Pennsylvania bank barns. They are incorrectly called Dutch barns, the true Dutch barns were built in New York state. The word German, when translated into German, is "Deutsch" and it easily got confused with "Dutch." Many barn historians call early German bank barns Sweitzer or Swisser barns.

The early barn builders were often German conservative Protestants, called the Plain People, which included the Amish, the Mennonites, and the Brethren. They brought traditions of a strong work ethic, a simple lifestyle, and good farming practices. They didn't decorate their barns. However, their less strict countrymen, the Lutheran and Reform, frequently decorated their barns with unusual artwork such as hex signs, fancy cupolas, and unusual brick work in the gable end walls.

The two distinguishing features of the German bank barn are the forebay and the earth bank. The barn has two stories with the upper story overhanging the first; the overhang is called the forebay. The barn is built into the side of a hill and the earth banked up to form a ramp leading up to the second floor — thus the name "bank" barn. A team of horses or a tractor can pull a wagon loaded with grain up the ramp to the second floor that has a threshing area and hay mows. On the first floor, or perhaps more correctly called the basement, are the stables. While the English built bank barns from New England to the Midwest, they didn't build them with a forebay.

The German bank barn was efficient since hay and grain could be carried to the second floor, stored, and later dropped through a chute to the livestock below. Straw from threshed grain could also be easily dropped to the barnyard below. The barns were big for their time — they were about 32 to 40 feet wide by 56 to 100 feet long. Large strong floors, built to give farmers of the 1700s space for grain threshing, are used by today's farmer to store heavy combines and other large farm equipment.

Origins

Prototypes of the German bank barn are found in most parts of alpine Europe (Germany, Switzerland, and Austria) with the strongest prototypes are found in eastern and central Switzerland. The best examples are in the Canton of Grunbunder. Authorities don't fully agree on the extent the German bank barn is an American invention of German immigrants as opposed to a European prototype brought to the United States. The prototypes discussed above are strong evidence that their origin is in Europe.

Swiss and German settlers came to southeastern Pennsylvania in the late 1600s bringing a strong tradition of animal husbandry. While other ethnic groups tended not to shelter their animals, the Swiss and the Germans did. They were skilled farmers and believed they could control feeding and breeding better by sheltering their livestock. Their barns, consequently, were not only designed for threshing grains, but also for animal husbandry.

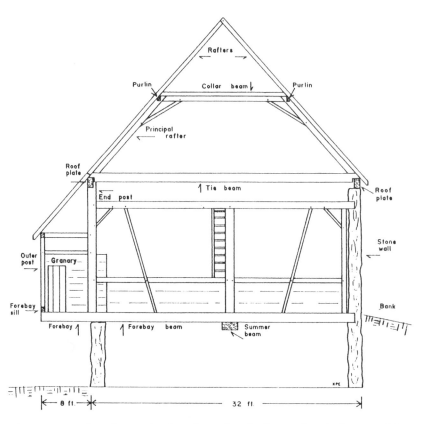

This cross-section shows the framing of early Pennsylvania barn, with a steep asymmetrical roof (sometimes called the Sweitzer barn). (Source: Robert Ensminger)

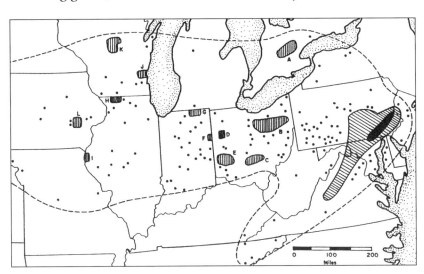

This map illustrates the distribution of the Pennsylvania barn. The solid black area is the core of intense concentration. The cross-hatched areas are major Pennsylvania barn regions. (Source: Robert Ensminger)

The core area where Germans and Swiss settled and built their bank barns was about 60 miles west of Philadelphia, centered in Lancaster, York, and Berks counties. The core area where these barns were built was about 30 miles wide by 120 miles long. From its center, the area ran southwest to the Maryland line and northeast to the New Jersey border.

The first German barns in Pennsylvania were log crib ground level barns. Two level barns with forebays appeared about 1730. Around 1750, German farmers started using heavy wood framing. For this period, these barns were quite large — often 40 by 100 feet. They had steep, 45-degree thatched roofs. The overhang or forebays were 6 to 9 feet. Usually the basement and gabled walls were masonry and heavy timber framing was used for the second floor walls and the roof. The roof line was asymmetrical — that is, there

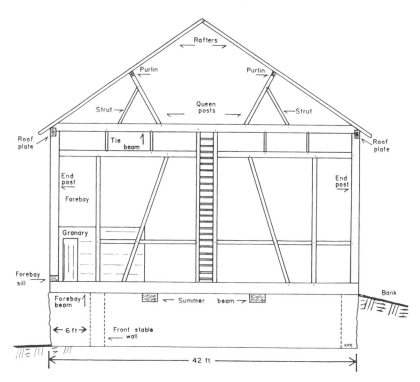

The forebays of early Pennsylvania barns appear as an "after thought," although they were built as an integral part of the original building.

The cross-section of this standard Pennsylvania barn shows a lower pitched, symmetrical roof. (Source: Robert Ensminger)

was a longer roof slope over the forebay side. While the forebay rested on cantilevered beams, the forebay wall and roof framing were, in appearance, an add-on to the main framing.

After 1790, farmers started building the standard German or Pennsylvania bank barn. The roof slopes were lower, 30-35 degrees instead of 45 degrees, because wood shingles replaced thatching. They built symmetrical roofs and the forebay became an integral part of the main framing. However, forebays were shorter, from 4 to 9 feet. Some theorize this change was because of the shortage of long timbers to make longer cantilevered beams. The first of the standard Pennsylvania barns were built with what is known as a closed forebay. The gabled end walls extended to the end of the forebay.

This large Pennsylvania barn near Lenhartsville, Pennsylvania, has a symmetrical roof.

This upper level floor plan of a standard Pennsylvania barn has two bays for threshing grain. (Source: Robert Ensminger)

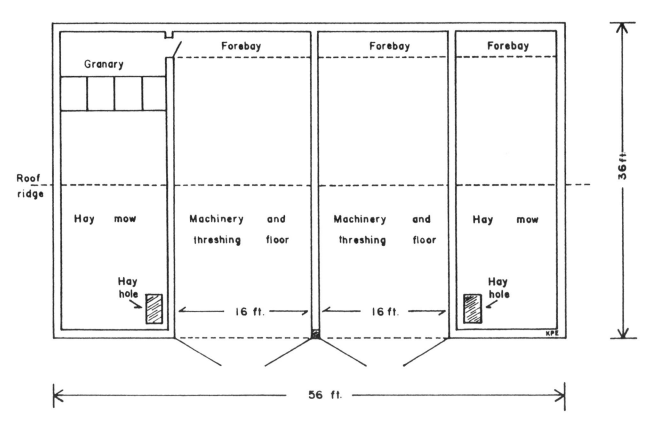

The threshing floor was on the second floor. It was in the center of the building and on either side were mows for hay, straw, and grain storage. Below, on the first floor, there were various arrangements of pens to house the cattle and horses and sometimes sheep, pigs, and chickens.

Later, because of increased agricultural production, farmers needed larger barns and started extending and enlarging them. For example, they extended the forebays from 15 to 20 feet. With these larger forebays it was necessary to support the ends with columns, often stone posts. Farmers also added front sheds as addition to the forebays and later added rear sheds on the bank side.

Construction

Pennsylvania bank barn builders would start by laying up the masonry basement walls and building a floor on top of the walls. The barn framing could be constructed in two ways: one, the timber posts, beams, rafters, etc., could be erected piece by piece as the framing went up; or two, the major timber members could be put together on the barn floor and then the assembled frame raised into place. A large crew raised the frame by pulling on ropes tied to the framing, while others pushed the framing up with pikes. The earlier barns tended to be assembled piece by piece, a European tradition. After 1860, farmers were articulating the framing on the barn floor and then neighbors helped erect the assembled frames. This new method became the American barn raising tradition.

The way the post, roof plate, and tie beam were connected was one factor that determined how the framing was put together. Before 1860, roof plates were connected under the tie beams. Later, builders connected the roof plate on top of the tie beam. They found this arrangement better for assembling and raising the timber framing as a whole.

Barns Spread Westward

The Pennsylvania barn soon spread from its core area to other sections of Pennsylvania and to other states. In the early 1700s the Germans and the Scotch/Irish took the barn into the Cumberland Valley of Pennsylvania, the Hagerstown Valley of Maryland, and down into the Shenandoah Valley of Virginia. After the revolution, Pennsylvania settlers moved into Ohio taking their barn building techniques with them.

As forebays got longer they had to be supported, instead of cantilevered. The forebay of this barn near Brinklow in Montgomery County, Maryland, is supported by stone columns. An open shed has also been added to the forebay.

This stone barn with ramp sheds is located near Dickerson in Montgomery County, Maryland. The barn was used as headquarters for the Union Army during the Civil War.

Many barn builders were the Plain People. They followed the Cumberland Trail or the National Highway west into Ohio. Some diverted southward on the Zanes Trace to Zanesville and Sommerset, Ohio. Many had been pacifists during the Revolutionary War and were not eligible for Veterans land benefits. After 1850, farmers were building Pennsylvania bank barns in states like Iowa, Nebraska, and Texas; eventually the Pennsylvania barns were built in Washington and Oregon.

Few Pennsylvania barns were built after 1920. The development of conveyors, blowers, and other material handling machines made the bank ramps unnecessary. While the barns were efficient, the below-grade basements were often subject to condensation and mildew. This development, along with the discovery of germs, and new-found interest in sanitation, ventilation, and light in the early 1900s, helped to bring the Pennsylvania bank barn to its end. Dr. Leonard Pearson, a public health reformer in 1885 who fought against bovine tuberculosis, condemned the bank barn as unsanitary. "The most cheerless place on many a farm is the basement of the old barn — dark, damp, and forbidding. Ceilings and walls that were covered with cobwebs and dirt and any disturbance of the floor above brought a shower of dust and hayseed" (Fletcher, p. 67). This perception of the bank barn as unsanitary was overblown; in fact with proper operator care and good ventilation, milk produced in a bank barn can be as safe as other types of barns.

The Amish and Mennonites, often using volunteer labor, still build and restore bank barns. Builders take older barns apart and reassemble them in new locations. Pennsylvania bank barns have been converted to riding stables and to shelter specialty livestock such as llamas, ostriches, and buffalo.

FROM THE DUST OF THE GROUND

Early Southwestern adobe ranch buildings stood in stark contrast to the Pennsylvania bank barn. For starters, the sunny, hot, dry, tree-sparse Southwest was so different from the cool, humid, dense Northeastern woodlands. Adobe builders and ranchers of the Southwest spoke a more romantic language, dressed more colorfully, danced more, worshiped with more symbols and liturgy, and had darker skin, eyes, and hair than the so-called Plain People of Pennsylvania and the Midwest. Spanish and Mexican ranches were huge — measured in square leagues that equal 4,300 acres. The low lying adobe buildings with flat roofs were built around a courtyard

and were in glaring contrast to the steep-pitched roofs of wooden Pennsylvania barns. In the Southwestern ranch complex, the house and livestock shelters were connected while the Pennsylvania bank barn was separate from the house.

Environmentalists appreciate the benign character of adobe buildings; they are built out of earth and return to earth as they deteriorate. Adobe construction has attracted artists like Georgia O'Keeffe to New Mexico where they painted the graceful earth-colored natural walls. Anthropologists and historians relish studying Southwestern adobe structures that are a mix of Pueblo and Spanish technologies because of their unique designs and what they reveal about these ancient civilizations. Southwestern Indians have built adobe structures since 1200 A.D. by puddling layers of wet mud.

The Spanish moved north across the Rio Grande Valley into what is now New Mexico in the mid-1500s. They built with logs, stone, and adobe. One early building was the jacal, a small primitive brush hut. It was first used for human habitation but was later used for animal shelters. The Spanish built these structures by knitting together small branches and plastering them with wet clay. This technique has been likened to the way a bird builds its nest. While the Spanish brought small saws and other woodworking tools, their use was mostly limited to cabinet and other furniture making. The Spanish didn't erect sawed lumber buildings until much later. However, the Spanish knew adobe brick-making technology, a technique that probably came from the Arabs or Moors when they occupied Spain. The word adobe comes from the Arabic word "atob" meaning sun-dried brick. Soon they were building fine adobe ranch buildings, villages, missions, forts, and presidios — much of it with Indian labor.

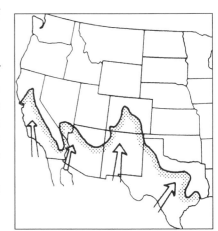

This map shows the extent of Hispanic settlements in southwestern United States. The Spanish built with adobe in most of this area, except in eastern Texas and northern higher elevations of Colorado and California. The Mormons also built with adobe in southern Utah. (Source: Noble, Vol. I)

La Hacienda

As cattle ranching and grain production prospered, the Spanish, and later the Mexicans, developed rural estates called haciendas. Later the word "hacienda" was used to refer to the main house on the estate. The really large estates were limited primarily to Mexico and southern California. The casa-corral on the hacienda was usually a large building called the casa mejor. It was often three sided, and was built around a courtyard. The hacienda had bedrooms, a dining room, a parlor, and kitchens, and sometimes a dance hall, music

A jacal, or palisade and daub building, was restructured at the Pioneer Arizona Living Museum north of Phoenix, Arizona.

Casa-corral

The casa-corral was a one-story farm building complex in which the casa (house) and corral were connected. The house, which was either rectangular, "L" shaped, or like a squared off "U," had five or six rooms built around a courtyard. Early casa-corrals had only a few small windows that were covered with shutters. Doors were low and small. Religious icons, pictures or statues depicting the Virgin Mary, the Saints, and the Holy family adorned the interior walls. The rooms, along with 4- to 8-feet high adobe walls, enclosed the courtyard. Behind and attached to the house's courtyard was another courtyard or corral for the livestock. Buildings for sheltering cows and horses and for storing hay and grain surrounded the corral. The casa-corral might be located on a few acres in a village, with the owner perhaps having additional land in outlying areas which village members shared as common pasture for cattle and horses.

This floor plan of Hacienda Martinez was the typical arrangement with two placitas or courtyards. The front courtyard was enclosed by rooms used for the family; the rear courtyard was enclosed with rooms or sheds for the servants, animals, and shops. (Source: Kit Carson Museums, Inc.)

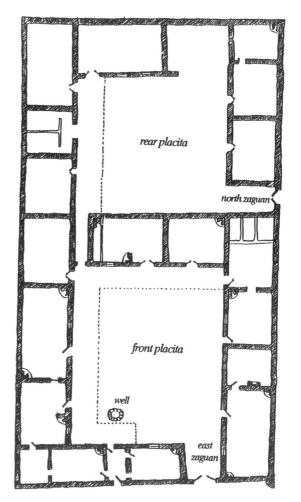

rear placita

north zaguan

front placita

well

east zaguan

The front placita (courtyard) of El Rancho de las Golondrinas near Santa Fe, New Mexico, is pictured here. The ranch was founded in early 1700s. On the right are the hornos (ovens).

room, billiard room, chapel, school, and wine cellar. Larger ranch houses may have had as many as two dozen rooms and a courtyard as large as 100 feet square. The open courtyard acted as a basin for cool air and may have had a drinking well, a windmill, and irrigated gardens with fruit trees and grape vines. The courtyard was enclosed by a wall across the back of the compound.

Right behind this back wall was a second courtyard with farm buildings around it to form another three-sided structure — often a kind of mirror image of the house. This structure contained the corral and the facilities necessary to run a 10,000-acre or larger ranch with 5,000 cattle or more. Within the corral, ranchers housed horses to pull the carriages, to ride on the range, and to work the fields. Calves, milk cows, prize bulls, breeding stock, and chickens were also kept in the corral. The corral buildings included workshops, stores, ranch hand quarters, the caballerisa for horses and tack, the tasolera — a building that stored hay above and dairy cattle below, and the barbareoa or granero for storing corn. There may also have been the troja that held the cookware and clothing. Most livestock housing consisted of sheds without side walls that faced

Looking from the front courtyard of Las Golondrinas Hacienda one sees the rear courtyard where the animals were sheltered.

the courtyard. They provided shade from the blistering southwest sun. Fleas and lice plagued the families who lived in the houses attached to livestock corrals.

The casa mejor had one or two covered entries (called el zaguan) with massive gates, which were large enough for carriages or large two-wheel wood carts. The complex was designed as a fort to prevent the Apaches or other Indian tribes from stampeding the cattle or to stop rustlers from stealing them. Other buildings, such as hay barns and grist mills, were outside the walls. Sometimes there were additional corrals or fenced cattle yards built of adobe, logs, or using mesquite posts interwoven with branches. Many cattle roamed the open range land.

This view of the rear courtyard of Las Golondrinas Hacienda shows fences and sheds made of log and adobe.

The granero (granary) of the Hacienda Martinez has adobe storage bins. (Permission of Kit Carson Museums Inc.)

Life at the Hacienda

In the morning the owner mounted a well-groomed horse with a western saddle and rode out to check his workers, the cattle, crops, and irrigation systems. The ranch was a feudal system employing hundreds of workers: supervisors, stable hands, a blacksmith, field hands, cooks, and sometimes a priest and medical doctor. Several ethnic groups lived and worked at the hacienda: Hispanics, Anglos, Indians, Negroes, and various mixes of these. Nobility, politicians, and church officials arrived from time to time to celebrate holidays and fiestas.

Fences were often made of adobe walls topped with branches of mesquite or other wood of the arid regions.

Forming and Building with Adobe Bricks

Ranchers built their hacienda buildings of adobe brick made of earth and straw. Ideally good adobe bricks should be an exact, well-proportioned mixture of sand and clay — with too much clay there would be shrinkage and cracks, and with too much sand the bricks would crumble. In practice, early ranchers tried to find a spot near the building site where the soil condition was reasonably well proportioned. They made adobe mud in shallow pits where they mixed earth and water by driving cattle through it. The mixed mud was packed in molds, forming from 1 to 30 large bricks which were usually 4 inches thick, 8 to 12 inches wide, and 12 to 16 inches long. The bricks would be removed from the molds, dried in the sun for a few days, and then stacked and dried for another two weeks.

The dried bricks, which weighed from 25 to 30 pounds, were laid up to make walls 18 inches to 3 feet thick. The walls were sometimes laid directly on the ground, but better constructed walls were built on a rock foundation. The foundation prevented moisture from deteriorating the lower portion of the wall. Builders usually tapered the walls, making them larger at the bottom and smaller at the top. Walls could be as high as 16 feet. Wood framing for the doors and windows was set in place as the wall was laid up. They often covered adobe walls with a mud plaster using their hands and not a trowel. Women often did the plastering. Adobe ranged in color from light orange or tan, to brown, gray, and black depending on the soil of the region. The interior of the dwellings would be plastered with a mixture of gypsum, wheat flour, and water. In later years exterior adobe was plastered with cement mortar because it helped the walls stand up longer.

Early ranchers constructed the floors of packed dirt. Later the more wealthy ranchers laid down log plank floors. The roof was made of wood poles that had the bark skinned off, called vigas. The poles were placed about 3 feet apart and usually protruded through the adobe walls. The vigas were only strong enough to span about 15 feet so this limited the width of a room to that dimension. Spanning the vigas were peeled saplings known as latias, and the latias were covered with smaller branches, reeds, or grass which were topped with clay and packed earth. In the house, ranchers covered the ceilings with muslin or other cloth to prevent grit from seeping from the roof. The roof may have a slight pitch towards the canales — water spouts that carried the water off the roof and away from the walls. The roof had a low parapet, often with gun ports. In the spring the prickly pears planted on the roof would bloom; in the fall, bright red chile peppers were hung to dry on the protruding vigas.

The problem with adobe was it became wind scarred, and rains and freezing weather could destroy it. If the roof wasn't rebuilt frequently, serious leaks occurred which damaged the walls. On the other hand, these mud buildings were easy to repair, and if kept plastered, could last many years. The farm buildings often weren't plastered and there are few good remains of hacienda corrals, barns, and livestock sheds.

The End of the Hacienda

After Mexico gained its independence from Spain in 1821 the Mexican government didn't have enough military power to patrol the frontier north of the Rio Grande. There were Indian raids, bandits, and conflicts with the Anglos, and the cattle industry nearly collapsed in the Southwest. Many haciendas were abandoned and left to deteriorate.

After 1821 the Santa Fe trail from Independence, Missouri, became operational and more Anglos moved into New Mexico and neighboring states, and fewer Mexicans came from Chihuahua,

The canales or water spouts of the Martinez Hacienda in Taos, New Mexico, carry the roof water away from the front wall.

The smooth adobe walls enclosed the rear courtyard of the Martinez Hacienda in Taos, New Mexico. (Permission of Kit Carson Museums Inc.)

Mexico. With the 1848 secession of the Southwest to the United States by Mexico, the end of the Civil War, and the control of hostile Indians, Anglo-Americans moved into the Southwest in even greater numbers. The cattle industry was revived. Anglo immigrants brought sawmills, window glass, and a new architecture. They still used adobe but built separate buildings instead of connected ones. They used larger windows and pitched gabled roofs covered with wood planks or sheet metal. The pitched roof created an outward thrust on the adobe walls and tie beams had to be added to keep the walls from being pushed over.

After 1891 the cattle industry declined in the Southwest and the railroad provided easy access to lumber. As a result, this colorful way of life and the hacienda soon ended. However, today there are still a few restored or replications of haciendas; good examples are El Rancho de las Golondrinas in Santa Fe, New Mexico, and los Martinez Hacienda in Taos, New Mexico.

SEEKING PERFECTION IN THE ROUND

"Thoughtless men go on building rectangular barns," wrote Wilber Fraser, an early 20th-century round barn crusader (Fraser, p. 14). Round barn crusaders were frequently zealots or eccentrics driven by a search for religious or economic perfection. Besides true round barns, reformers experimented with barns having from 6 to 20 sides. From a distance they looked round and are called non-orthogonal barns — that is, barns without square corners. Some corn cribs in the Midwest look round but are actually two semicircular structures attached to a square. The square area is the driveway and the two semicircular areas are corn bins.

Shakers constructed a large stone round barn in Hancock, Massachusetts, in 1826. They are known for their fine furniture and buildings that they built to the glory of God. Circles such as sewing,

This elegant stone Shaker round barn is near Pittsfield, Massachusetts. (Source: Hancock Shaker Village, Pittsfield, Massachusetts. E. Fitzsimmon, photographer)

George Washington designed and built this 16-sided threshing barn which has been restored and can be seen today at Mount Vernon.

George Washington, Father of the Round Barn?

He wasn't a zealot nor an eccentric, but George Washington was a clever innovator who designed and built one of the first round-like barns near his home in Mount Vernon, Virginia, in 1793. It was a 16-sided, two-story structure for threshing grain. The barn was nestled against a hill allowing wagons to enter the upper floor where the grain was threshed. The floor, made of heavy wood planks with cracks between them, allowed the grain to fall to the floor below, leaving the chaff above. The round shape fit the circular pattern that the oxen or mules walked when threshing grain. Round barns didn't catch on at the time and only a few wealthy gentleman farmers built them. The restoration of this barn can be seen today at Mount Vernon.

singing, and praying were an important part of Shaker life and were prevalent in their designs. Old traditions held that circular structures were good because they didn't have corners where the devil could hide. In many religions, the circle represents the heavenly and the square the worldly. The circle can be found in many Eastern Orthodox icons. And today in secular terms, people sometimes refer to an unsophisticated, dull person as "a square." So whether it's religion or the secular, the circle is considered good and the square or rectangle bad. While the Shaker barn was quite elegant with many windows and a large cupola, this 90-foot diameter barn was not very efficient. It held fewer than 54 cattle and was viewed as an oddity by neighbors. The round barn did not catch on with the common farmer.

In the latter half of the 19th century, more secular reform movements drove the demand for round barns. Orson Squire, an eccentric health reformer and a strong phrenology advocate, promoted the octagonal home for health and better living. Scientists at land grant colleges and agricultural experiment stations were searching for better farming methods and facilities. Carpenters were building with lighter wood framing which could be formed into circular walls. The time was ripe for round barn champions.

Many round and round-like barns were built between 1860 and 1920. Before 1890 mainly octagonal barns were built; after that, innovators started building the true round barn. Most round barns were built in the Midwest: Wisconsin, Iowa, Illinois, Ohio, Nebraska, and Indiana. Probably every state in the Union has at least one round barn and several were built in Oregon and Washington. For the most part, farmers who built round barns were better educated and more innovative. These barns were more complex and complicated to build but had several advantages over conventional rectangular barns.

Convenience, Strength, and Cheapness

Wilber Fraser claimed round barns had three points of superiority over the rectangular one: convenience, strength, and cheapness. A growing interest in convenience was evident around the turn of the 20th century. This interest came from the scientific management movement. New terminology was being added to the vocabulary: industrial engineering, work sampling, time and motion study, and efficiency experts. These all led to an interest in better industrial plant layouts and designs. Frederick Taylor and the husband and wife team of Frank and Lillian Gilbreth, leaders in the movement,

Round Barn Advocates

- *Elliot W. Stewart*, a farmer, agricultural scientist, and lecturer at Cornell University, built an octagonal barn in New York in 1878. He wrote several articles in agricultural journals proclaiming its benefits.

- *Franklin King*, an agricultural physicist at the University of Wisconsin Agricultural Experiment Station, worked on the design of silos and round barns. He built a true round barn with a center silo for his brother near White Water, Wisconsin, in 1889.

- *Wilber Fraser*, a dairy specialist at the University of Illinois, published two bulletins, one in 1910 and another in 1918, advocating round barns in the strongest terms.

- *Orson Squire Fowler*, a mid-19th century reformer, phrenologist, sex educator, and eccentric, is credited with promoting octagonal building. He also promoted dress reform, vegetarianism, and water cures.

were contemporaries of Fraser. It's reasonable to expect that they influenced his thinking. In any event, round barn proponents like Fraser pointed out that the animals could face a center feeding station such as a silo or grain bin, reducing the distance the farmer had to travel to feed the livestock. Baby pigs could also huddle close to a central heater in small octagonal barns.

In terms of strength, round barns were streamlined and could better resist high winds from tornadoes and hurricanes than the flat walls of a rectangular barn. Advocates compared the round barn to an indestructible barrel and a traditional barn to a flimsy box. Some even called the round barn tornado proof. While in theory round barns might better resist tornadoes, it would be an exaggeration to call them tornado "proof."

As far as cheapness, a round barn has a smaller wall perimeter than a rectangular barn with the same floor area. This meant fewer materials to buy, less wall area to paint and maintain, and fewer linear feet of foundation to build. The crusaders used some simple but powerful mathematical formulas to compare the relationship between floor area and wall perimeter. These are the same formulas one learned in high school geometry, i.e., the area of a circle equals π times the diameter squared over four, and the circumference equals π times the diameter. In theory for any given area, a square has 13% more perimeter than a circle of the same area. A rectangle with a length twice its width, has close to 20% more perimeter than a circle of the same area. Wilber Fraser claimed construction savings as much as 58% for round barns. However, in this analysis, he compared round barns to rectangular ones that were almost five times longer that they were wide.

Diameters: Less than 15 Feet to Over 120 Feet

Most round barns had diameters of 50 to 70 feet. However, there was some variation. Small octagonal barns with diameters less than 15 feet were used for pig houses; while others were much larger — there is a round barn with a diameter of 122 feet in Franklin County, Iowa. Many round barns were built on two levels with a ramp leading to the second level, like the Pennsylvania bank barn. The second level had a hay mow and below were the stalls for cows and horses. Some barns also had large pens holding loose cattle; many were built just for dairy cattle.

In later true round barns, builders laid out the pens in a circle. Sixty foot diameter barns had one ring of stalls with the cattle facing the barn's center toward a silo or grain bin. Around the silo or grain bin, in front of the animals, was a feed alley, and behind the animals was a litter alley. This layout held between 35 and 40 cows, or about 10 horses and fewer cows. Larger 90-foot diameter barns had two rings of stalls having cows facing each other with a feed alley between them. Behind each ring of animals was a litter alley. These larger barns could hold up to 100 cows: 65 in the larger outer ring and 35 in the inner ring. In the center was often a silo that might be 20 feet in diameter. Some layouts didn't have a silo, but had a driveway through the barn

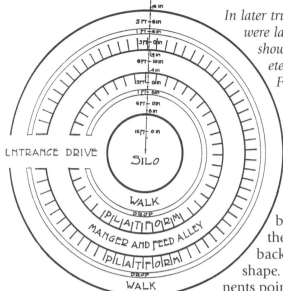

The early octagonal barns used a rectangular layout for the stalls and pens, and didn't take advantage of the circular shape of the barn.

In later true round barns, stalls were laid out in a circle as shown in this 90-foot diameter barn. (Source: Wilber Fraser)

which allowed the farmer to drive a wagon into the barn for delivering hay and other supplies.

Stalls in circular barns were narrow in the front and wide in the back — a trapezoidal shape. Round barn proponents pointed out that when the cow laid down, her hips were wider than her head and so these irregular shaped pens were ideally shaped for cattle.

It Took a Special Builder

Round barns were built throughout the Midwest by itinerant builders with special round barn skills. Sometimes local builders who had picked up the skill constructed a few round barns in their area. Another source was precut round barns listed in farm catalogs.

Wood balloon framing of a round barn being built in Gratiot County, Michigan, in 1905. (Source: Gratiot County Historical and Genealogical Society, Jefferson Arnold, photographer)

The roof framing for the Gratiot County, Michigan round barn creates spectacular patterns. (Source: Gratiot County Historical and Genealogical Society, Jefferson Arnold, photographer)

Builders started by laying a round concrete foundation and basement wall of brick, stone, or hollow clay blocks. It wasn't hard to make a circular wall with these materials. Some terra cotta plants made clay blocks with a curved face that were made especially for round barns and silos. On top of the basement wall carpenters built the wood framing. Light lumber balloon framing was becoming prevalent and it was easier to shape into circular patterns than heavy timber construction. Carpenters framed the main walls of 2- by 6-inch or 2- by 8-inch lumber, spaced 2 feet 6 inches on center. The barn siding usually consisted of boards laid horizontally or vertically; when laid horizontally, they were bent to conform to the barn's curvature.

The hay mow floor was typically framed with 2- by 12-inch joists that pointed toward the barn's center. They were spaced 2 feet 6 inches on center at the outer perimeter and were much closer at the barn's center. The flooring was 1-inch wooden planks laid in different patterns over the joists with a double layer for driveways. The basement floors were usually concrete.

Round barn roofs looked like large hats or large upside-down bowls. They were conical shaped with one pitch or they had two or three different pitches. The latter were called gambrel roofs. The roof framing was made of 2- by 6-inch or 2- by 8-inch rafters. Carpenters learned to build self-supporting cone or gambrel roofs providing an unobstructed hay mow without posts for smaller barns. The roof was supported by only the outer walls. Larger barns required extra posts for the roofs. The previously mentioned 122-foot diameter barn had a flat roof carried by wire cables attached to the top of a silo.

One unique requirement of the round barn was the special design for the plate supporting the roof. This was the ring on top of the stud wall that had to resist the outer thrust caused by the weight of the roof and snow loads. The roof plate was often made with six 1- by 6-inch members 10 feet long, placed on edge. Lumber would be wet or green so it could be curved to conform to the circular wall. A similar arrangement was used for the sill plate and the ring supporting the roof cupola.

Claims Exceed Practice

Farmers built few round barns after 1920. The crusaders had vigorously promoted round structures with both mysticism and science but had exaggerated their benefits and farmers found the round barn impractical. Another reform movement, better sanita-

This snow-covered round barn is in good condition near Centre Hall, Pennsylvania. (Source: Dennis Buffington)

This stone round barn, used for horse training, was built about 1975 near Kittanning, Pennsylvania.

While only a fraction of a percentage of the barns ever built in this country were round, many of them are in historic preservation programs. People seem fascinated with round barns and many have been maintained in splendid condition. Some have been restored as restaurants, gift shops, homes, and stores. In the 1950s and '60s, semicircular and circular corrals were built around a central milking parlor in California. They were built on the basis that the milking parlor would be central and convenient to the cows holding pen. While this arrangement has not caught on, perhaps we have not seen the end of the round barn. New technologies could bring it back; what goes around comes around.

tion, helped bring the round barn to an end. The new reform encouraged more light and ventilation. Long, narrow, rectangular barns with many windows provided good cross-ventilation and lots of light — round barns didn't. They were difficult to enlarge and many carpenters were not adept at building them. While in theory the round barn used fewer materials, lumber had to be cut at an angle, wasting much of it. With the introduction of baled hay, square bales didn't fit well in round barns. The introduction of the internal combustion engine tractor required barns with wide, straight aisles for the tractors to operate effectively. New material handling equipment such as belts, screw conveyors, and blowers also didn't work well in a round configuration.

This round dairy barn in Champaign, Illinois, has been converted into a restaurant.

MODERN COW CITY

"The health and even the lives of the children of this country depend to a great extent upon the conditions existing in the barns where milk is produced." (Eckles and Anthony, p. 533). These words, of a noted early 20th century dairy scientist, Clarence H. Eckles, are perhaps even truer as we move into the 21st century. One modern dairy now supplies milk for tens of thousands of children. John Mullady, another dairy specialist writing in 1853, claimed 8,000 to 9,000 children a year die of diseases caused by impure milk (Fletcher, p. 197). The modern dairy operation is not only sanitary but big, efficient, and tied more to commerce than to the land. Before discussing the modern dairy barn, let's briefly trace a few of the landmark events that changed the dairy barn from a single cow pen to freestall housing with thousands of cows.

During the late 19th and early 20th centuries, industrial firms started manufacturing automatic dairy feeding and watering systems, plus steel stanchions and stall dividers. The University of Wisconsin provided plans for dairy barns and silos. Mechanical refrigeration and pasteurization helped provide a safer milk supply. Commercial dairy farms developed with herds of 25 to 30 cows. One family plus one or two hired hands could milk a herd this size. Farmers were building brighter and cleaner dairy barns.

The cool weather and rich soils of the Great Lakes regions provided perfect conditions for growing lush grasses and legumes

This dairy barn in Somerset County, Maine was photographed in 1936. (Source: National Archives)

Early Milk Delivery

In the middle of the 19th century, and even into the 20th century, town and city dwellers had their own cow or bought milk from a neighbor who had one. Around 1840 the first trains brought milk to New York City and Boston in wooden churn-shaped containers from farms 50 to 60 miles away. Soon small dairy farms grew up near Northeastern cities where "milk runs" brought fresh milk to the city dwellers in the familiar 10-gallon metal milk cans which had replaced the earlier wooden ones. Milk dealers met the trains early in the morning and delivered fresh milk door to door by a horse-drawn milk wagon carrying the milk cans from which the housewife would dip milk using her own pitcher.

for hay, pasture, and silage. The highest quality forage was grown here. The big cities of New York, Chicago, Philadelphia, and Detroit demanded huge quantities of milk. The eight states that border the Great Lakes — New York, Pennsylvania, Michigan, Wisconsin, Minnesota, and the northern parts of Ohio, Indiana, and Illinois — became the major milk-producing region. At least half the milk produced in this country was produced in these states.

The rural landscape in the Great Lakes region and other states of the Northeast and Midwest became dotted with long barns with large gambrel or arched roofs with tall vertical silos attached. This is the barn most think of when one mentions dairy barns. They were two-story, above ground buildings with the cows on the first floor and hay in a large loft above. The truss roof framing had no cross ties and provided an unobstructed clear span for a hay distribution track. A hood at the roof ridge protected a hay loft door. Rows of windows lined each side of the barn providing ample lighting and ventilation. These barns housed two rows of cows kept in gray steel stanchions or tied up in small stalls where the farmer milked them. Usually the cow faced outward, and there was an alley behind them

In the early 20th century there was a growing interest in improving sanitation through better ventilation and light as in this Douglas County, Nevada barn. (Source: National Archives)

where a manure spreader could be driven through the barn. Manure was shoveled from the gutters behind the cow into the spreader. Pens were at one end of the barn for maternity cows and young stock. These barns, unlike the round barns, were easy to lengthen to accommodate an expanding herd. The older barns where cows were kept in a dark basement soon gave way to this more efficient and sanitary barn.

While the milking machine was coming into use in the 1920s and '30s, most farmers still milked by hand and sent the milk to market in metal cans. Although much of the milk production was in the Great Lakes region, every state in the Union produced milk. Even until World War II, most farmers were milking one or two cows by hand. This provided enough milk for the family and livestock, and enough cream to sell at the local general store.

After World War II, commercial herds became larger — sometimes with over 100 cows. As people moved southward and westward, so did dairy production — moving into southern and central California, Florida, and Texas. Although farmers couldn't grow high quality forages in many of these areas and cows were subject to heat stress, the areas offered other advantages. Dairy farms were near new milk markets, and there was an abundant supply of cheap immigrant labor. Hay could be shipped in from states like Colorado or in the Southwest it could be grown locally with irrigation, and it cost less to build dairy housing in these milder climates.

Dairy farmers were replacing the traditional two-story gambrel roof dairy barn and vertical silo with low, one-story loose and freestall housing with horizontal silos. They were building low-cost pressure-treated pole structures with prefabricated wood roof trusses and sheet metal or wood panel siding. Cows were no longer kept in stanchions or pens but were free to move about in the barn.

In 1963, a Mr. Dugan from Wisconsin moved to Arizona for his health. He started dairying with only a dozen or so cows. Now his sons, the Dugan brothers, have their own dairies near Casa Grande, Arizona, each with thousands of cows.

Dairy farms continue to be less dependent on soil fertility, because they don't have to be operated in the cool climates where the lush forage grows. Environmental regulations on manure disposal have become more important. Biologically engineered hormones may help boost production. Larger and better insulated or refrigerated tank trucks enable milk to be shipped hundreds of miles. As a result, entrepreneurs are clearing the deserts in Nevada, Arizona, and New Mexico of mesquite trees and creosote bushes to build huge modern dairies. In these mild dry climates, manure is easier to handle and dispose of; odors are less of a problem; and lower cost shelters can be built. In the Northwest, environmental regulations near growing urban centers have restrained dairy farming in western Washington, while production in remote areas of eastern Washington and Idaho has been increasing. Dairy farmers have established large dairies in selected remote areas of the prairie states — Kansas, Nebraska, South Dakota, and in the mountain states of Colorado and Utah.

More farmers were using milking machines and milking parlors. Milking parlors were first used in the 1930s and had a pit where operators could stand with their hands and arms at the level of the cow's udder so they could easily attach the milking machines. The first parlors were small, handling two to four cows at a time. The 1,000-gallon milk tank and tank trucks were replacing the familiar metal milk cans.

Larger Herds = Larger Dairies

As the 20th century comes to a close, dairies have become even larger. Herds of 300 to 600 cows are typical with some herds as large as 10,000 cows. The cows are kept in open lots or in freestall housing. The larger dairies are highly automated and are often owned and operated by large family corporations.

The large modern dairy farm has housing for cows, a feed center, waste disposal areas, a milking center, a machine shop, and equipment storage. Many farms have housing for calves, but some larger operators sell all their calves and buy replacement stock from special breeders. They find this approach is more economical, they don't have to care for the very young calves which is the most labor intensive and they usually get better bred calves.

In the mild dry climates of Arizona, New Mexico, Nevada, and southern California, cows are housed in open dry lots with the cows segregated by stage of lactation. The only shelters are sun shades. These shades are as large as 100 feet wide by a quarter mile long. Since there is little or no snow load, they are built of light steel framing. The structures have flat or low-pitched metal roofs and no sidewalls. They are equipped with fans and water misters or sprinklers to protect the cows from the burning southwestern summer sun. The buildings have either concrete or dirt floors.

In Florida and other parts of the Southeast where it is warm but wet, dairy operators built similar lightweight buildings with no sidewalls. However, they use railing around the open sides to control the cows. Water runoff from lots or pastures often contaminates the ground water and regulations require farmers to keep cows confined in buildings for better waste disposal control.

In colder areas, like Nebraska and Colorado, cows are housed in enclosed freestall barns. The barns are long, low, narrow buildings, 90 to 120 feet wide, as long as 1,200 feet, and may hold 1,000 cows. They are built of heavy steel framing or wood truss framing to resist the heavy snow loads. The roof is usually low pitched and covered with sheet metal. The sides are 12 to 14 feet high and have plastic curtains that can be rolled up and down. The body heat of the animals can usually help maintain the building at adequate temperatures. In the coldest climates and for buildings housing calves, the sidewalls and roof are insulated and the building housing calves is provided with auxiliary heat.

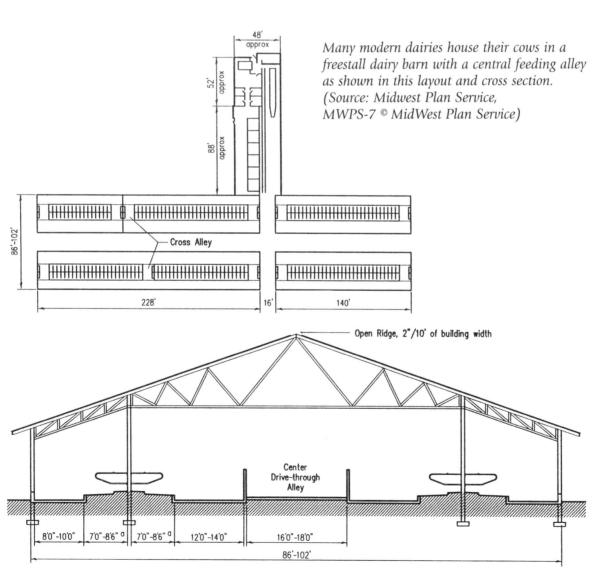

Many modern dairies house their cows in a freestall dairy barn with a central feeding alley as shown in this layout and cross section. (Source: Midwest Plan Service, MWPS-7 © MidWest Plan Service)

Long open-sided shelters are used at this dairy near Stanfield, Arizona. Note the large fans near the roof eaves.

Fan and misters (misters not shown in this photo) helped keep the cows cool at the dairy near Casa Grande, Arizona.

Freestall barns have stalls about 4 feet wide and about 8 feet long where cows can lie down. The stalls have railings on each side and in the front, but the back is open so the cow is free to move in and out of the stall. Cows can be confined to the barn or be free to move out into dry lots or pastures. The larger barns have four rows of stalls with the cows either facing each other or placed back to back. There are various arrangements of manure and feed alleys servicing the cows. Most stall floors are concrete and covered with a bedding of straw, peanut hulls, sand, or wood chips. Sometimes rubber mats or mattresses filled with bulk materials are used.

Cows need only minimum shelter in the mild climate at this dairy near Hemet in southern California.

Hay storage and feeding alley at dairy farm near Hemet in southern California.

The Feed Center

The modern dairy has a large feed center where the grain, hay, protein supplements, and silage are stored. In the Southwest, hay bales are stacked in open sheds without side walls. Sometimes it is just stacked in the open with no protection from the weather. In wetter climates the hay may be stored in enclosed buildings. Silage is stored in horizontal 100-foot long plastic bags, piled on asphalt or concrete pavement where it is covered with plastic sheets, or stored in vertical silos. Corn, barley, cotton seed, and mineral protein supplements are kept in bins, sheds, or tanks, sometimes called commodity barns. Many modern dairy farms use a mobile unit on the back of a truck or tractor for weighing and mixing feed. The silage, grain, or hay is moved from their bins or stacks to the mobile unit by front end loaders or augers. Here it is weighed and mixed to provide the best feed ration. The truck or tractor carries the feed from the feed center to the dry lots or the freestall barns.

Many barns, no matter how the cows are arranged or the type of shelter, have a center feeding alley. It is 16 to 18 feet wide and a feed truck is driven up and down the alley to distribute feed. In the dry lots, the feed truck delivers the feed to bunkers along the fence row.

Manure Removal

In the open lots of the Southwest, manure is cleaned up with a front end loader several times a year. It is loaded on a truck and piled in stacks at the back of the dairy. The dairy operator may deliver and spread the manure for other farmers, or other farmers may pick up the manure with their own equipment. The manure in enclosed freestall barns is removed by automatic scrapers on an endless chain or flushed out with water. Some systems have slatted concrete floors in the manure alley allowing the manure to fall into a pit below.

Plastic bags store silage at a dairy farm near Casa Grande, Arizona.

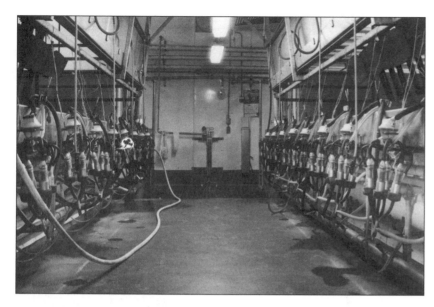

Lowell Dell's fine milking parlor near Beatrice, Nebraska, has two rows of eight milking machines.

Fifty cows line each side of the T&K Red River Dairy's immaculately clean milking parlor near Stanfield, Arizona.

The Milking Center

At milking time groups of cows move to the holding area of the milking center — this is the heart of the dairy farm. Besides the holding area, the milking center includes the milking parlor, the office, the milk room with bulk storage tanks, a utility room, a lounge, and rest rooms.

In the milking parlor, workers stand in an operator pit that is about 3 feet below the level where the cows stand. The operators clean the udder and attach the milking machines. Most modern pits are 6 feet and wider, and some are over 100 feet long. There are many different arrangements for the parlor; in one arrangement the cows stand behind one another and parallel to the length of the pit; another arrangement is called the herringbone and the cow stands at a 30° angle to the pit. In the more popular one for large herds, the cows stand parallel to each other and perpendicular to the pit. In the modern large parlor, as many as 50 cows may be on each side of the pit — for a total of 100 cows. With a parlor this size, operators can milk as many as 600 cows per hour. Milking is a well choreographed operation with cows, workers, and automatic machinery all moving in concert.

From the milking parlor, the milk flows into glass or stainless steel pipes through a heat exchanger that reduces the temperature of the milk from 97° to 37°. The cooled milk then flows into large steel holding tanks. Each tank may hold as many as 8,000 gallons. Milk from the tank is pumped into large tanker trucks and delivered to milk processing plants.

Most of the milking centers are one-story buildings constructed of concrete block. The interior walls are faced with tile or coated with epoxy to give a smooth sanitary surface. Some are built of steel or wood framing with the interior wall and ceilings paneled with steel, aluminum, or plastic. These modern milking centers are immaculately clean, well lighted, and well ventilated. They remind one of hospital facilities or a scientific laboratory. Some larger farms are divided into modular units of 1,500 to 2,000 cows to each unit with its own milking parlor.

Today, modern large dairy production facilities are producing more milk with better cows. Farmers are producing a safe and wholesome food and protecting the health and lives of American families.

BARNSCAPES

VIEWED FROM AFAR

"A monstrous perversion of good taste" (Allen, p. 13 and 14). These words about traditional red barns were written by Lewis Allen, a mid-19th-century architectural critic and reformer. He further wrote that farm architecture was "offensive to the eye," and believed farmers had a moral duty to build well-designed buildings. Beauty must be in the "eye of the beholder," because many photographers, painters, and sketch artists, and many non-artists, see beauty in the old 19th-century barns. Maybe time has changed these "monstrous perversions" into objects of charm, quaintness, and beauty.

In any event, artists like to include barns in their pictures because they accent a sometimes monotonous landscape. Fences around the barnyard tie the buildings together or direct the eye around the picture until it reaches a focal point. The barn often is balanced by several smaller outbuildings. Barns seem to belong on the landscape showing mankind's benign intervention on nature. Artists see barns and farmsteads differently than most folks and differently from each other. Often many artists convey feelings and attitudes by painting pictures filled with distortions and exaggerations.

Grant Wood was an early 20th-century painter who painted in Iowa and other parts of the Midwest. His most famous painting, "American Gothic," showed a strident farm couple with the man holding a pitchfork in front of a Gothic farmhouse. One of his best farmscapes is "Fall Ploughing," a somewhat abstract painting of a storybook farm with a plow in the foreground, big round trees, and salt shaker shaped grain shocks. The plowed field and row of shocks lead the eye to the barn in the background. The scene looks as

The barn and silo are viewed from a distance in this Grant Wood painting, "Fall Plowing." The scene has an artificial and make-believe expression. (Source: Deere & Company)

HOME TO THANKSGIVING.

Rural life and farmsteads were popular scenes during the 19th century such as this Currier & Ives engraving, "Home to Thanksgiving."
(Source: Library of Congress)

though it were a model built from hobby store materials. His rolling hills seem exaggerated for Iowa and could just as well be hills in Pennsylvania, North Carolina, Oregon, Washington, or New York.

The farmstead in "Home to Thanksgiving," one of Currier & Ives famous lithographs, does not have the sense of distance as Grant Wood's painting. The snow-covered house and barn, the people on the porch, the worker in the barn, the horse and sleigh, the cattle, and the dog all seem in their proper places and in order. It's quaint, charming, and sentimental, yet maybe even a little stilted, too old fashioned. While the main action, the hosts greeting their guests, is on the farmhouse porch, the barn is prominent.

BARNS ACROSS AMERICA

Rural life and farmstead scenes were popular during the 19th century for greeting cards, calendars, advertisements, and pictures for the living room wall. The barn frequently had a conspicuous place in the picture, which may include a farmhouse, smokehouse, spring house, crib, and shops and other farm things like wagons, cattle lots, fences, and windmills. And like "Home In Thanksgiving," it may include people. Currier & Ives mass produced their pictures by lithography. They used part-time artists drawing on stones with wax crayons. The picture was printed in black and white, then enhanced with color by hand on an assembly line.

Contrast the Courier & Ives painting and Wood's painting to Grandma Moses' painting "In Harvest Time." Grandma Moses was a self-taught 20th-century primitive or folk painter who painted around Albany, New York. Her colors are lighter and brighter than Grant Wood's painting and the large gray barn and the white house in the foreground dominate the scene. Grandma Moses includes people and many activities in her picture; it's busy — with workers harvesting grain and men doing business — it is the most striking difference between the other two paintings.

Currier & Ives, Moses, and Wood's work are all cheerful in comparison to Andrew Wyeth's somber paintings. He was a contemporary of Grandma Moses who painted in the Brandywine Valley of Pennsylvania and in Maine. He used browns, ochre, orange browns, off whites, and grays. Wyeth's farmscapes denote isolation and loneliness although many were painted within 30 miles of downtown Philadelphia. He usually did not show people or activities around the farm and only minimal tools, equipment, or animals. He sometimes included only one person or one piece of equipment which even increased the sense of isolation. One of his

Andrew Wyeth paints somber farmsteads as seen in "Christina's World." (Source: The Museum of Modern Art, New York. Purchase. Photograph © 1997 The Museum of Modern Art, New York)

most famous paintings, "Christina's World," portrays the isolation of a Maine farmstead. It shows Christina, a crippled woman, crawling toward a farmhouse and barn at the top of the hill. Wyeth saw the farm as a utilitarian, lonely, tough workplace, while Moses saw it as something pleasing, active, and cheerful.

Many barns stand alone on Western ranches. Even Grandma Moses couldn't miss the loneliness of a barn setting isolated in a vast panoramic western landscape with no other buildings or people in sight. The opposite of the solitary scene is the congested and overcrowded scene where urban sprawl, institutions, and facto-ries crowd out barns. The barn no longer fits; it also seems isolated here.

Each part of the country has its own unique barnscape. In New England, the salt box barns often appear with a slanted roof facing north or toward the winds. Barns and other farm buildings were connected to, or located near, the house so the farmer didn't have to go outside much in the severe winters. In the mid-Atlantic states, large bank barns were set into the hillside. In the South, farm buildings tended to be small and separate buildings with small barns and separate cribs. In the Midwest, the farmsteads were laid out on flat land and in straight patterns. The buildings were set parallel to the roads that ran north and south or east and west. These neat patterns were the result of the flat lands and go back to the township system of land surveys that set up farm sections one mile square. Some midwestern barns lie in low flat land subject to

An isolated abandoned building in southern Utah now stores irrigation equipment.

A seldom-used isolated barn is located near the Tetons in western Wyoming. (Source: U.S. Department of Agriculture, 1975 photo)

This lone barn is near Pine Creek, Oregon. (Source: Ilene Wellman)

flooding. In the Southwest, low profile adobe structures seemed to rise out of the ground. The walls were the same color as the soil. Western high deserts and mountains offered dramatic settings for the barn. The warm colors of the barn siding often were in stark contrast to the snow covered mountains.

Barnscapes change by the season. In the spring, the barn is surrounded by plowed fields, blooming trees, and planting activities. In summer, crops are at their peak and the blooming wildflowers surround the barn. Summer barnscape paintings are filled with splashes of warm tones — like red, yellow, and orange. In the fall, the barnscape is filled with bright fall foliage and harvesting activities. In the winter the snow reduces the landscape — it is quiet and still. Eric Sloane, a 20th-century artist who sketched many barns, believes barns are best seen in the winter. The bleached gray timbers become part of the winter landscape. Impressionist painters exaggerated the blues in the winter barnscapes.

In the mid-20th century, agricultural engineers, soil and water conservationists, and efficiency experts developed more efficient farmstead layouts and better farm buildings. Plan Services and Extension agencies, operated by land grant colleges and U.S. Department of Agriculture, distributed blueprints and other information on farmsteads and building designs. They promoted layouts where the house, barn, other outbuildings, the farm lots, and fields were oriented in the best way to reduce labor. New building materials such as cement blocks, plastics, and sheet metal replaced wood and stone. Barnscapes changed. Efficiency sometimes replaced charm. On the other hand some of the new soil and water conservation practices enhanced barnscapes. Farm ponds were added to the barnscape; they served as reflecting pools heightening the beauty of the farmstead. Farmers planted hay and small grain between contoured strips of corn and other row crops. This created broad alternating green and gold ribbons sweeping across the rolling hillsides.

Downtown Richmond, Virginia, is within sight of this dairy barn and silos. (Source: U.S. Department of Agriculture, 1980 photo)

Modern buildings surround this solitary dairy barn and silo on Texas Tech.'s campus in Lubbock.

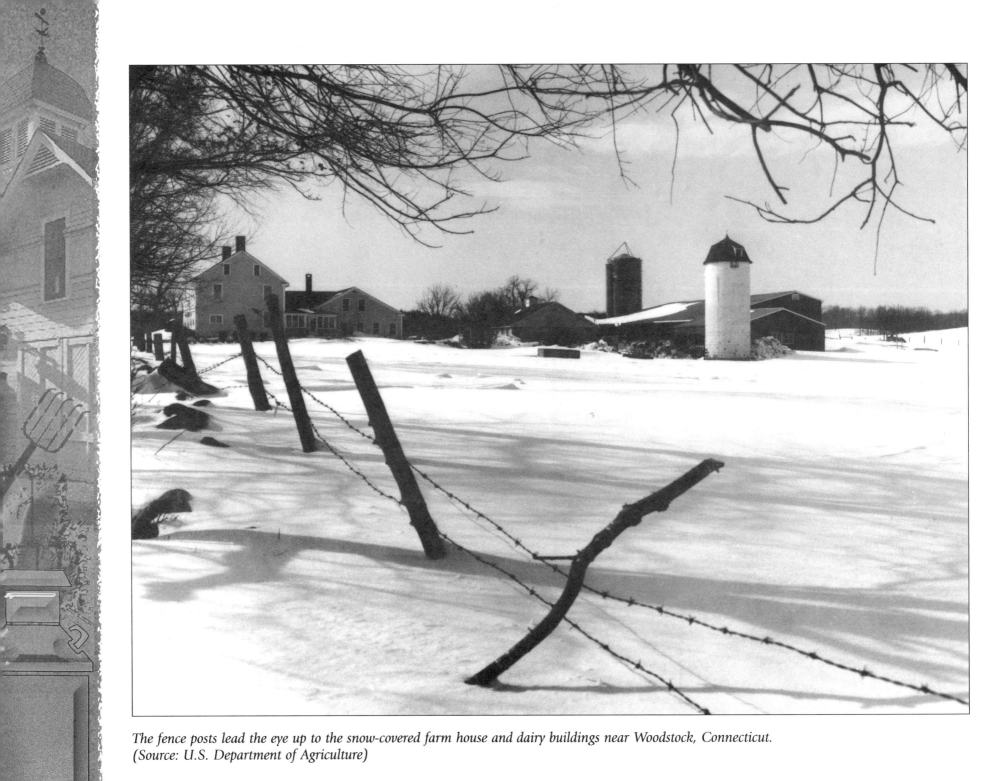

The fence posts lead the eye up to the snow-covered farm house and dairy buildings near Woodstock, Connecticut.
(Source: U.S. Department of Agriculture)

The barns and silos are nestled into a rolling Carroll County, Maryland landscape. (U.S. Department of Agriculture)

In the Midwest, roads and farm buildings are lined up according to compass directions. (Source: U.S. Department of Agriculture)

The white trim sets off the roof lines of the crib barn in southeastern Nebraska, near Wilber.

This farm is near the Sierra Nevada Mountains near Carson City. (Source: Whittington Collection, University of Southern California, Neg. J1-34-8)

This deserted farm is located in the high Escalante Desert of southwest Utah.

The dark, low, broad roof frames a stark white three-portal barn near Ogallala, in western Nebraska.

The 1962 spring rains flooded this sheep farm near Charlesville, Minnesota. (Source: U.S. Department of Agriculture)

In springtime in central Maryland, the barn is set off with the dogwoods in bloom, lush green pastures, and white-board fences. The Montgomery County government has designated this lane as a Rustic Road.

This barn on Maryland's Eastern Shore rests in a sea of summer flowers. (Source: James T. Walker)

Fall barnscapes are often favorites of photographers. This one is located near Woodstock, Vermont. (Source: Eugene H. Lambert)

A decaying barn near Glenwood, Washington, as seen in the winter with Mount Adams in the background. Since this picture was taken the barn has fallen down and the rubble burned. See page 121 for a picture of this same barn taken in the fall. (Source: M.D. Locke, Jr.)

The first snow blankets this West Virginia barnscape. (Source: James T. Walker)

The pond reflects the image of this Virginia barn. (Source: James T. Walker)

A summer southeastern Washington farmstead displays contoured alternating strips of green lentils and yellow wheat.
(Source: Eugene H. Lambert)

A Closer Look

Georgia O'Keeffe painted close-ups of barns. The barn's roof ridge almost touched the top of her canvas, and the base of the barn was near the canvas' bottom. In her painting, "Lake George Barns," there are no contoured fields, ponds, farm animals, trees, children playing, or other outbuildings, but the canvas is filled with massive barn walls. With barns like this, it's no wonder we have the expression, "He couldn't hit the side of a barn."

Are her buildings well proportioned? Architects over the century have developed rules for well-proportioned buildings. Ancient Greeks had mathematical formulas that architects used to design beautiful, appealing, and harmonious buildings. One famous rule was the golden rectangle. This rule postulated that perfect order and beauty were achieved where the ratio of the height to the width was 0.625. This approximates the proportions of a 3- by 5-inch index card. The front of the famous Greek Parthenon uses these proportions. Renaissance architects used a system of parallel diagonal lines, called regulating lines, to proportion the width and height of rectangular windows and doors. On the other hand, most farmers and frontier carpenters didn't use mathematical relationships when designing barns. They built their barns to accommodate the hay, the cattle, the grain, and the wagon. Louis Sullivan, a great 19th-century

Georgia O'Keeffe liked to get in close when she painted barns as in her "Lake George Barns" (1926). (Source: Collection Walker Art Center, Minneapolis. Gift of the T.B. Walker Foundation, 1954)

architect and teacher of Frank Lloyd Wright, said, "Form follows function" (Winter, p. 50). He meant that if a building is designed to meet its purpose it will look good.

While the barns O'Keeffe painted may not have been designed by traditional rules of proportion, they appealed to her aesthetic senses. She described most barns as "beautiful, but many of the farm houses look like bad accidents" (Robinson, p. 19). Barns looked right in spite of their large disproportionate shape. While bulky and awkward, they have a certain appeal like a big awkward elephant or an early American large wooden clothes chest.

Like Georgia O'Keeffe, this photographer has gotten a close-up of this plain, but charming, 1898 red barn near Hampton Township, Illinois. (Source: Rock Island Historical Society. Gene Dennhart, Photographer)

cupolas made the barn more interesting. They could be aesthetically pleasing when designed to meet a function.

Good barn ventilation was needed to keep the barn cool in the summer and prevent condensation in the winter. To meet this need farmers often created attractive decorative patterns with their ventilation system. Farmers built openings in the brick or stone barn end walls. These might be in the form of simple slots (called loopholes) or in complex patterns that looked like wheat shocks, stars, trees, or flowers. They designed decorative openings in the wood siding or built long narrow louvers in the walls, often trimming them in white. These ventilation openings were sometimes forms of creative expressions for the farmer designing the barns.

Farmers installed cupolas with louvers on top of the roof for added ventilation which often enhanced the barn's beauty. Another roof form was the weather vane. They had been used for years and were one of the first barn decorations. While decorating barns was considered frivolous and vulgar by early American strict religious groups, the weather vane was allowed because it supposedly served a function.

By the late 1800s, carpenters were using the band saw and jig saw to create elaborate gingerbread decorations. These were especially used in carriage houses and stables and called Victorian designs.

As one moves closer to the barn, one observes the texture in the barn materials. Hand plastered adobe has a special cool wavy texture. The most common barn material, wood aged by weather

Some barns, though, built by amateurs, are quite graceful and proportioned. Many find the beauty of the barn enhanced by the various contrasting roof lines, silos, and sheds and other additions.

While many barns, especially early ones, were simple and plain and had a certain elegant minimalism, over time designs became more elaborate. Moving through the 19th-century, barns were designed with more windows, bigger doors, cupolas, louvers, weather vanes, lightning rods, and fancy hardware. Early barns had few if any windows, but by the mid-1800s windows were larger and more frequently used. The number of window panes often varied with the region of the country. The addition of windows and

This barn has a long sweeping lean-to roof on one side against a background of cedars and other evergreens, near Glenwood, Washington. (Source: M.D. Locke, Jr.)

and wind, has a mellow patina not duplicated in artificial materials. Some smooth modern manufactured materials come across as cheap and bogus in comparison. Wood is considered a sympathetic material because it doesn't present a shock of cold or heat. The weathered wood texture is enhanced when combined with the texture of a stone foundation or end walls. Stone without mortar has an especially pleasing textural effect.

A backdrop of bright yellow Lombardi poplar trees shows off this white gothic roof barn near West Linn, Oregon. (Source: M.D. Locke, Jr.)

This cattle barn, with the large sweeping gable roof near Glenwood, Washington, is representative of some Pacific Northwestern barns built at the turn of the 20th century. (Source: M.D. Locke, Jr.)

This barn and corral were built in the late 19th century near the foot of the Grand Tetons in Wyoming. (Source: John Telford)

This barn shows off some unique planes, angles, and corners. It was painted black, like many barns in central Kentucky, giving it a sense of mystery.

This cable-on-hip-roof looks like a massive hat on this tobacco barn. It was built before the Civil War and now stands on a public golf course near Mitchellville in Prince George's County, Maryland.

The low profile and windowless plain adobe walls of Martinez Hacienda near Taos, New Mexico, are quite a contrast to conventional American farm buildings.

This hand-plastered hacienda adobe wall offers a wavy cool texture. (Permission of Kit Carson Museums Inc.)

The windows of this historic limestone barn in Gage County in southeast Nebraska, built by master craftspeople, look as if they belong in a cathedral.

Basement windows and doors of a bank barn in Berks County, Pennsylvania.

A "storybook" red barn with a gambrel roof and trimmed in white is located near Mollalla, Oregon.
(Source: M.D. Locke, Jr.)

The white trimmed windows and the white doors are a striking contrast to the dark red paint of this Maryland barn located only 20 miles from the U.S. capital.

Unusual decorations are incorporated in the gable end wall brickwork of this giant barn at Emmitsburg, Maryland.

The long narrow vertical vents and the high cupola make this barn in Somerset County, Pennsylvania, look like a large Gothic cathedral.

This carriage house, built in 1887, has an elegant cupola and is located near downtown Redlands, in southern California.

The artist sketch of the same carriage house looks a little more informal than the photograph. (Source: Earl Thollander)

The wood roof shingles present a fine warm texture on this old tobacco barn in Prince George's County, Maryland.

This old barn in Loudon County, Virginia, has a graceful roof and fine textured wood after years of disrepair. (Source: U.S. Department of Agriculture)

Colors and Symbols

Early barns were not painted. Painting was considered extravagant, vulgar, and showy, and many farmers couldn't afford it. In the mid-20th century, many barns in the South and Appalachia were still not painted. Even if a barn has been painted once, it is often not repainted but instead is left to fade over time. Unpainted, faded, bleached weathered barn timbers have a certain esthetic appeal. Eric Sloane, a 20th-century barn artist and writer, says there is a certain beauty of wood "in a state of pleasing decay" with moss and lichens. He calls this one of "Nature's special masterpieces" (Sloan, *Age of Barns*, p. 13). Products stored in the barn such as hay, straw, tobacco, and corn add subtle mellow colorful tones to the barn.

In the early 19th century, northern and mid-Atlantic farmers started painting their barns a dark red. Several theories exist on why red was such a popular color for barns. Some barn authorities claim it was the Scandinavian influence; they used red to simulate brick and wealth. Others say it was an abundance of stock blood or iron oxide that could be mixed with milk to make red paint. Others suggested it was esthetics — the red paint complimented the green fields. Yet another theory suggests it was a supply and demand tradition. Farmers, when asked why they painted their barns red, replied, "red paint is so available and cheap." If paint manufacturers asked why they produced so much red paint, they said, "because so many farmers want it." In any event, a rich dark red has become the symbolic barn color in America.

Unpainted and fading wood siding and shingles add to the beauty of this barn near Virginia City, Montana. (Source: David Orcutt)

Other colors are also used on barns. White is another popular barn color, along with gray, blue, and green. In the counties of central Kentucky, such as Mercer County, barns and fences and other farm buildings are painted black. Sometimes they are trimmed in white but are often painted completely black. This color scheme came from the tradition of using lamp black and diesel fuel as a cheap wood preservative. Old adobe barns of the Southwest were earthen color.

The Pennsylvania barns used hex signs for decorations, and contrary to popular belief, were seldom used to ward off evil spirits. Haciendas in the Southwest had religious symbols and icons, for example, paintings and statues of the Holy Family.

This fall scene of the gray timbers and wood shingles of this decaying barn near Glenwood, Washington, is made more colorful by the yellow lichens. (See page 100 for this same barn in a winter scene.) (Source: M.D. Locke, Jr.)

The Grand Tetons offer a dramatic background to this barn with gray and brown unpainted siding and shingles. (Source: John L. Telford)

Mellow yellow drying tobacco hangs in a Maryland barn. (Source: James T. Walker)

There is a unique attractiveness to this unpainted wood barn door with the rusting hinge. (Source: David Orcutt)

The red buildings are in stark contrast to the shiny metal bins and the white fences at this farm in Woodhull, Illinois.

Many barns have been painted red with white trim like this one at Frying Pan Park in Herndon, Virginia.

This barn and silo near Glenwood in southeast Washington are painted a deep rich red. (Source: Eugene H. Lambert)

The blue truck stands out among the bright white farm buildings in Frying Pan Park, Herndon, Virginia.

Fences, cupolas, and barn walls show a lot of white at the Sotterly Plantation in St. Mary's County, Maryland.

The green barn walls match the field grass surrounding this barn located on the Gilmen Historic Ranch in Banning, California. (Source: Cate Whitemore, Curator of History, Riverside County Historical Commission Photographic Collection)

This light beige barn is located in Montgomery County, Maryland, near Poolesville.

The American flag decorates the side of this barn near Des Moines, Iowa. (Source: U.S. Department of Agriculture)

celebrity (Jackson, p. 13). He painted Mail Pouch signs in West Virginia, Pennsylvania, and Ohio for 45 years — he estimates he painted 20,000 signs. Mail Pouch barn signs were also painted all over the Midwest and in California, Washington, and Oregon.

Firms like Mail Pouch sent out advertising agents, or sometimes just the sign painter, to approach farmers for permission to advertise on their barns. Agents searched for barns that could be easily seen from the road, that were not hidden by trees and bushes, and that were in good condition. They didn't want dilapidated barns that would reflect poorly on their product. The farmer was coached with a variety of incentives to get the required permission: agreements to repair barn roof leaks, paint the entire barn, a gold watch, coupons for free tobacco or the other products that were to be advertised, or cash — if necessary. In recent years some barn owners have paid to have Mail Pouch ads painted on their barns; these ads have become a kind of heirloom.

Painters used ladders and scaffolding to paint the signs often working freehand from small-scale drawings. They tried to paint at least one sign a day even though the thick lead-based paint required constant stirring and took a long time to dry. But the durable lead-based paint had a life of 30 to 40 years. Some painted with it have bled through later signs painted with a less durable product. These are like an aberration, showing the original sign; other old barn signs are just fading away. These "ghost signs" reveal a past commerce in chewing tobacco and other popular products for the times.

This thermometer was advertising Coca Cola on a barn in Frying Pan Park in Herndon, Virginia.

Barn Advertising

"Chew Mail Pouch," "Jesus Saves," or "John Deere Farm Machinery," are a few of the signs which have been painted on barn walls and roofs. Patriotic, political, religious, and especially commercial advertising on barns was popular from 1890 to 1950, a time when people traveled mostly on two-lane country roads where barns had a commanding presence. After 1950, barn advertising gave way to electric signs and mega billboards on well-traveled, high-speed interstate highways.

Ads for cigarettes, chewing tobacco, and bulk tobacco were particularly prominent on barns, with signs for Bull Durham, Mail Pouch, and Prince Albert. Mail Pouch barn ads were the most widespread. The company, based in Wheeling, West Virginia, started painting ads on barns in 1897 and didn't discontinue it until 1992. Harley Warrick was a Mail Pouch painter who has become a minor

This West Virginia barn carries the famous Mail Pouch ad. (Source: James T. Walker)

This barn near Ephrata, Pennsylvania, carried this design in the early 1940s. (Source: Library of Congress)

This barn near Fayetteville, Arkansas, is referred by locals as the "Jesus Barn." It shows years of Christian witnessing. Ghost signs of previous paintings have bled through later coats of paint. (Source: Arkansas Historic Preservation Program)

Classical sculptures have supplemented the old-fashioned hex decor on this barn near New Smithville, Pennsylvania.

Light comes through the door and cracks in the wall of a tobacco barn near Upper Marlboro in Prince George's, Maryland.

INSIDE THE BARN

As one moves inside the barn, there are other interesting patterns and feelings that emerge. Earl Thollander who sketched barns in California said, "I like barns. There is something good about entering their dark and cavernous, airy interior" (Thollander, preface) As one enters an old barn, the first impression is the lack of light. The dim daylight gives barns a sense of mystery and enigmatic beauty. The bright lights from open doors and large windows provide a stark contrast between the barn's dark interior and the outside world. Splashes of concentrated light come from cracks in the walls, nail holes in the roof, and from small louvers. The play of lights and shadows helps one perceive the forms and textures of the barn's interior. The play of lights and shadows changes throughout the day and through the seasons.

In the interior of the barn one can observe interesting rhythmic patterns of rows of stalls, windows, louvers, or feeding devices. In modern barns and food factories, there is a continuous repetition of stalls and other equipment. This repetition may be boring to some, but others may find it extremely exhilarating and interesting.

Contrasting spaces of different sizes and heights provide excitement and surprises. This is particularly evident as one moves from the low headroom of basements and stairways to the lofty heights of the hay loft. Some barns, with large openings from the floor, through the hay loft, and up to the ridge of the barn roof, are compared to lofty cathedrals. These contrasting spaces are peeked by the lines of massive timbers that run vertically, horizontally, and even diagonally. The wood timbers, often joined by pegs and dowels, lend a certain organic nature to the barn.

Stacks of baled hay and huge obsolete farm machinery all enhance the excitement and mystery of the barn's interior. The barn is a world of many shapes — things long and thin, fat and short, round and triangular; it's a world of many materials: wood, cast-iron, burlap, canvas, leather, hemp, steel, or crockery. Andrew Wyeth, Earl Thollander, and Eric Sloane have all painted or sketched interiors of barns showing things that are found there — milk cans, bags of grain, tools, equipment, troughs, and bins.

Light from the cracks and the vent fall on the hay bails in this interior view of a hay loft of an old barn built in the early 1800s near Poolesville, Maryland. The small light beam on the upper left is coming through a loop-hole in the stone end wall.

Light invades this barn in Montgomery County, Maryland. Through wall cracks it streams across the floor and hay.

Light from an open door falls across the barn alleyway in Frying Pan Park, in Herndon, Virginia.

The doorway inside one farm building frames the end of this barn north of San Luis Obispo near the California coast. (Source: David Orcutt)

Light shines through cracks in the door and walls in this hay loft of the Stabler Barn in Montgomery County, Maryland.

Light streaks can be seen in the foreground on the adobe wall and in the background through the open archway at the Martinez Hacienda in Taos, New Mexico. (Permission of Kit Carson Museums Inc.)

This graceful archway is in a stone barn built in 1906 in Garfield County in north central Oklahoma.

Looking upward one sees a striking opening to the cupola in this octagonal barn near Grottoes in the Shenandoah Valley of Virginia.

The basements of many barns have low ceilings and dark small spaces like this one near Woodhull, Illinois.

Upstairs in the same barn, the space is high and wide for hay and grain storage.

136

BARNSCAPES

An expansive hay loft is in this Madera County barn in central California built in 1870. (Source: Earl Thollander)

Hay bales create warm textures and rhythmic patterns.

The pattern of the sturdy vertical timber posts of this Somerset County Maine barn is broken by the lone cow. (Source: National Archives)

BARNSCAPES

The metal stanchions of varying heights produce contra patterns on either side of the aisle of this dairy barn near Kalamazoo, Michigan. (Source: National Archives)

In modern pig houses like this one near Milford, Utah, the pattern of these feeding bottles goes on and on.

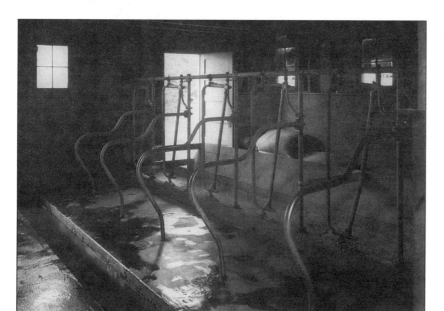

Graceful steel cow rails in a dairy barn at the Indian training school at American Forks, Utah. (Source: National Archives)

The new large food factories are replacing traditional barns. What new plastics or space-age materials will they be built of? These factories already house thousands of animals or chickens at one location and are built with elaborate feeding and waste disposal systems. Will tomorrow's artists draw, paint, or photograph the new biotech age food factories? What will happen after all the old barns are gone?

THE VANISHING BARNSCAPE

Georgia O'Keeffe, the renowned 20th-century artist, described the big red barn as a brooding hen that settled "calmly and absolutely into the gentle slopes of the field" (Robinson, p. 19). Early farmers seem to have had an innate sense of how to site their buildings on the land. In the old days one could ride down most any rural road and see many great barnscapes. And barnscapes were brought into the home in the form of paintings and wall calenders.

Today, one can still drive to the countryside and see pleasant barnscapes. But they are getting fewer and farther between. Barnscapes are an endangered specie. They are being taken over by residential subdivisions, industrial parks, and the new food factories. Calm barnscapes are needed more than ever as relief to today's overwhelming and nervous visual stimuli of roadsides filled with fast food restaurants and advertising blasts from TV, giant billboards, and magazines.

Concrete vats and clay pots were located in this olive curing shed built in 1912 on the Gilman Ranch in Banning, California.

Laufenberg Barn Wall

Most barns have a collection of tools and other miscellaneous
items, like this one in Sonoma County, north of San Francisco.
(Source: Earl Thollander)

Old barn machinery takes many
interesting forms and shapes.

REFERENCES

Note: Many of the references have been used for several chapters, but we have shown most references under the chapter of their major use. The last group of references are general books on barns that are used throughout the book.

Chapter 1

Ahler, Stan. *People of the Willows.* Grand Forks: University of North Dakota, 1991.

Allen, Lewis. *Rural Architecture: Farm Houses, Cottages, and Out Buildings.* New York: C. M. Saxton, 1852.

Ballantine, Betty and Jan Ballantine. *The Native Americans Illustrated History.* Atlanta, GA: Turner Publishing, Inc., 1993.

Bidwell, Percy W. and John I. Falconer. *History of Agriculture in the Northern United States 1620-1860.* New York: Peter Smith, 1941.

Brown, Ralph H. *Historical Geography of the U.S.* New York: Harcourt, Brace, and Company, 1948.

Chitwood, Oliver Perry. *A History of Colonial America.* New York: Harper and Row, 1961.

Cochrane, Willard W. *The Development of American Agriculture: A Historical Analysis, 2nd. ed.* Minneapolis, MN: University of Minnesota Press, 1993.

Condit, Carl W. *American Building Materials and Techniques from the First Colonial Settlements to the Present.* Chicago, IL: The University of Chicago Press, 1968.

Eberlein, Harold Donaldson. *The Architecture of Colonial America.* Boston: Little, Brown, and Company, 1929.

Ferris, Robert G., ed. *Prospector, Cowboy, and Sodbuster*, Vol. XI. Washington, DC: U.S. Department of Interior, National Park Service, 1967.

Fledderus, Mary L. and Mary van Kleeck. *Technology and Livelihood.* New York: Russell Sage Foundation, 1944.

Gowans, Alan. *Styles and Types of North American Architecture.* New York: Harper Collins, 1992.

Gray, Lewis Cecil. *History of Agriculture in the Southern United States to 1860, Vol. I & II.* Glouchester, MA: Peter Smith, 1958.

Grow, Lawrence. *Country Architecture.* Pittstown, NJ: The Main Street Press, 1985.

Hayden, Dolores. *Seven American Utopias.* Cambridge, MA: The MIT Press, 1976.

Lavender, David. *The American Heritage History of the Great West.* New York: Bonanza Books, 1965.

Hurt, R. Douglas. *American Agriculture A Brief History.* Ames, IA: Iowa State University Press, 1994.

Jennings, Jerry. *The West.* Grand Rapids, MI: The Fideler Company, 1979.

Jones, Maldwyn Allen. *American Immigration, Second Edition.* Chicago, IL: The University of Chicago Press, 1992.

Klamkin, Charles. *Barns — Their History, Preservation and Restoration.* New York: Bonanza Books, 1979.

Larkin, David, June Sprigg, and James Johnson. *Colonial Design in the New World.* New York: Stewart, Taburi, & Chang, 1988.

Lytle, R. J. and others. *Farm Builder's Handbook.* Farmington, MI: Structures Publishing Co., 1973.

Maxwell, James A. *America's Fascinating Indian Heritage.* Pleasantville, NY: The Readers Digest Association, Inc.,1978.

McGaw, Judith A. *Early American Technology: Making and Doing Things from the Colonial Era to 1850.* Chapel Hill, NC: University of North Carolina Press, 1994.

McMurry, Sally Anne. *Families and Farmhouses in Nineteenth-Century America.* New York: Oxford University Press, 1988.

Meinig, D. W. *The Shaping of America, Volume I.* New Haven, CT: Yale University Press, 1986.

Middleton, Richard. *Colonial America: A History, 1607-1760.* Cambridge, MA: Blackwell, 1992.

Morison, Samuel Elliott, et al. *The Growth of the American Republic, Volume II.* New York: Oxford University Press, 1969.

Morrison, Hugh. *Early American Architecture from the First Colonial Settlement to the National Period.* New York: Oxford University Press, 1952.

Novak, Philip. *The World's Wisdom.* San Francisco: Harper, 1995.

Partridge, Michael. *Farm Tools through the Ages.* Boston: New York Graphic Society, 1973.

Paterson , J. H. *North America: A Geography of the United States and Canada.* New York: Oxford University Press, 1994.

Preservation League of New York State. *Farmsteads and Market Towns.* Albany, NY, 1982.

Rasmussen, Wayne D., ed. *Readings in the History of American Agriculture.* Urbana, IL: University of Illinois Press, 1960.

———. *Agriculture in the United States, Vol 1.* New York: Random House, 1975.

Rebeck, Andrea. *Montgomery County in the Early Twentieth Century: A Study of Historical and Architectural Themes.* Silver Spring, MD: Montgomery County Historic Preservation Commission, 1987.

Roberts, Sam. *Who We Are: A Portrait of America.* New York: Random House, 1993.

Rounds, Glenn. *The Treeless Plains.* New York: Holiday House, 1967.

Sanford, Albert H. *The Story of Agriculture in the United States.* Boston: D.C. Heath and Co., 1916.

Sears, Roebuck & Co. Catalog, 1927 Ed. Alan Mirken, Editor, Bounty Books, Division of Crown Publishers, Inc., 1970.

Stilgoe, John R. *Common Landscape of America: 1580 to 1845.* New Haven, CT: Yale University Press, 1982.

Teter, Norman C. and Henry Giese. "New Barns for Old." *Power to Produce: The Yearbook of Agriculture 1960.* Washington, DC: USDA, 1960.

Todd, Lewis Paul and Merle Curti. *The Rise of American Nations, 2nd ed.* New York: Harcourt, Brace, and World Inc., 1950.

Tunis, Edwin. *Indians.* Cleveland, OH: The World Publishing Co., 1959.

Vlach, John Michael. *Back of the Big House: The Architecture of Plantation Slavery.* Chapel Hill, NC: The University of North Carolina Press, 1993.

Chapter 2

Borie, Beauveau IV. *Farming and Folk Society, Threshing Among Pennsylvania Germans.* Ann Arbor, MI: UMI Research Press, 1986.

Caras, Roger A. *A Perfect Harmony: The Intertwining Lives of Animals and Humans Throughout History.* New York: Simon & Schuster, 1996.

Carter, William. *Middle West Country.* Boston, MA: Houghton Mifflin Book Company, 1975.

Fletcher, Stevenson Whitcomb. *Pennsylvania Agriculture and Country Life, 1840-1940.* Pennsylvania Historical and Museum Commission, Harrisburg, 1955.

Franklin, Wayne, ed. *Reflecting A Prairie Town.* Iowa City: University of Iowa Press, 1994.

Little, Charles E., ed. *Louis Bromfied at Malabar: Writings on Farming and Country Life.* Baltimore, MD: Johns Hopkins University Press, 1988.

Greene, Hank. *Square and Folk Dancing.* New York: Harper and Row, 1984.

Danbom, David B. *Born in the Country: A History of Rural America.* Baltimore, MD: Johns Hopkins University Press, 1995.

Hasselbrock, Kenneth. *Rural Reminiscences: The Agony of Survival.* Ames, IA: Iowa State University Press, 1990.

Hastings, Scott E., Jr. and R. Elsie. *Up in the Morning Early: Vermont Families in the 30's.* Hanover and London: University Press of New England, 1992.

Hostetler, John A. *Amish Society, Forth Edition.* Baltimore, MD: Johns Hopkins University Press, 1993.

Hoy, Jim and Tom Isern. *Plains Folk: A Commonplace of the Great Plains.* University of Oklahoma Press, Norman and London, 1987.

Jager, Ronald. *Elegy for a Family Farm.* Boston: Beacon Press, 1990.

McCormick, Virginia E., ed. *Farm Wife, A Self Portrait, 1886-1896,* Ames, IA: Iowa State University Press, 1990.

Murphy, Jim. *The Great Fire.* New York: Scholastic Inc., 1995,

Neugebauer, Janet M., ed. "Diary of William DeLoach 1914-1964." *Plains Farmer.* College Station, TX: Texas A & M University Press, 1991.

Parsons, William T. *The Pennsylvania Dutch A Persistent Minority.* Boston, MA: Twayne Publishers, a Division of G. K. Hall & Co., 1976.

Porter, Dennis. "Dolly Douthitt." *Garfield County Historical Society's Book of History 1893-1982.* Topeka, KS: Josten's Publications, 1982.

Schreiber, William. *Our Amish Neighbors.* Chicago: The University of Chicago Press, 1962.

Smith, Gene. *American Gothic: The Story of America's Legendary Theatrical Family — Junius, Edwin, and John Wilkes Booth.* New York: Simon & Schuster, 1992.

White, E. B. *Charlotte's Web.* New York: Harper & Row, 1952.

White, E. B. *The Second Tree from the Corner.* New York: Harper & Row, 1954.

Winters, Donald L. *Tennessee Farming: Tennessee Farmers.* Knoxville, Tennessee: University Press, 1994.

Yale, Allen R., Jr. *While the Sun Shines: Making Hay in Vermont 1789-1990,* Montpelier, VT: Vermont Historical Society, 1991.

Chapter 3

Adams, Robert H. *The Architecture and Art of Early Hispanic Colorado.* Colorado Associated University Press, in cooperation with the State Historical Society of Colorado, 1974.

Bunting, Bainbridge. *Taos Adobes: Spanish Colonial and Territorial Architecture of the Taos Valley.* Albuquerque, NM: University of New Mexico Press, 1992.

Eckles, Clarence H. and Ernest L. Anthony. *Dairy Cattle and Milk Production.* New York: McMillan Co., 1956.

Ensminger, Robert F. *The Pennsylvania Barn: Its Origin, Evolution, and Distribution in North America.* Baltimore, MD: Johns Hopkins Press, 1992.

Fowler, Orsen S. *The Octagon House, A Home for All.* New York: Dover Publications, 1973 (originally published 1853).

Fraser, Wilber John. *Economy of the Round Dairy Barn.* Bulletin, University of Illinois Agricultural Experiment Station No. 143, Urbana-Champaign, 1910.

Hanou, John T. *A Round Indiana: Round Barns in the Hoosier State.* West Lafayette, IN: Purdue University Press, 1993.

Hutslar, Donald. *Log Construction in the Ohio Country 1750-1850.* Athens, OH: Ohio University Press, 1992.

Jordon, Terry G. *Texas Log Buildings: A Folk Architecture.* Austin, TX: University of Texas Press, 1978.

MidWest Plan Service (MWPS). *Dairy Freestall Housing & Equipment, MWPS-7, fifth edition.* Ames, IA: Iowa State University, 1995.

Moffett, Marian and Lawrence Wodehouse. *East Tennessee Cantilever Barns.* Knoxville, TN: The University of Tennessee Press, 1993.

Romero, Orlando and David Larkin. *Adobe Building and Living with Earth.* Boston, MA: Houghton Mifflin Co., 1994.

Sanford, Trent Elwood. *The Architecture of the Southwest: Indian, Spanish, American.* New York: W.W. Norton & Co., 1950.

Soike, Lowell J. *Without Right Angles: The Round Barn in Iowa.* Iowa City, IA: Penfield Press, 1990.

Spears, Beverly. *American Adobes: Rural Houses of North New Mexico.* Albuquerque, NM: University of New Mexico Press, 1986.

Stewart, Janet Ann. *Arizona Ranch Houses.* Tucson, AZ: The Arizona Historical Society, 1974.

Weslager, C. A. *The Log Cabin in America: From Pioneer Days to the Present.* New Brunswick, NJ: Rutgers University Press, 1969.

Chapter 4

Bayley, Stephen. *Taste the Secret Meaning of Things.* New York: Pantheon Books, 1991.

Brown, Richard W. *Pictures from the Country.* Charlotte, VT: Camden House Publishing, Inc., 1991.

Bry, Doris and Nicholas Callaway, eds. *Georgia O'Keeffe: The New York Years.* New York: Alfred Knopf, 1991.

Corn, Wanda. *The Art of Andrew Wyeth.* Boston, MA: New York Graphic Society, 1973.

Dijkstra, Bram. *America and Georgia O'Keeffe.* New York: Alfred Knopf, 1991.

Duff, James H. and others. *An American Vision: Three Generations of Wyeth Art.* Boston: Little Brown and Co., 1987.

Goldstein Ernest. *Grant Wood American Gothic.* Champaign, IL: Garrard Publishing Co., 1984.

Hicks, Roger. *35 mm Panorama.* New York: Sterling Publishing Co., 1987.

Jackson, Debby Sonis. "Mail Pouch Signs Today." *Golden Seal,* Winter 1994.

Kostof, Spiro. *America by Design.* New York: Oxford University Press, 1987.

Moses, Anna Mary Robertson. *Grandma Moses: My Life's History.* New York: Harper & Brothers Publishing, 1952.

Rasmussen, Steen Eiler. *Experiencing Architecture.* Cambridge, MA: The M.I.T. Press, 1959.

Rifkind, Carole. *A Field Guide to American Architecture.* New York: The New American Library, 1980.

Robinson, Roxana. *Georgia O'Keeffe.* New York: Harper Collins, 1989.

Sloane, Eric. *An Age of Barns.* New York: Ballantine Books, 1967.

———. *American Barns and Covered Bridges.* New York: Funk & Wagnell, 1954.

Stage, William. *Ghost Signs, Brick Wall Signs in America.* Cincinnati, OH: Sign of the TImes Publications, 1989.

Thollander, Earl. *Barns of California.* San Francisco, CA: California Historical Society, 1974.

Tucker, Pauls. *Monet in the '90s.* New Haven, CT: Yale University Press, 1989.

Winter, Nathan B. *Architecture is Elementary: Visual Thinking Through Architectural Concepts.* Salt Lake City, UT: Gibbs Smith Publisher, 1986.

General — Barns and Farm Buildings

Arthur, Eric and Dudley Whitney. *The Barn: A Vanishing Landscape in North America.* Greenwich, CT: New York Graphic Society Ltd., 1972.

Carter, Deane G. and W. A. Foster. *Farm Buildings, Third Ed.* New York: John Wiley & Sons, Inc., 1949.

Endersby, Elric, Alexander Greenwood, and David Larkin. *Barn: The Art of a Working Building.* Boston, MA: Houghton Mifflin Co., 1992.

Fink, Daniel. *Barns of Genesee County 1790-1915.* Geneseo, NY: James Brunner Publisher, 1987.

Fitchen, John. *The New World Dutch Barn: A Study of its Characteristics, Its Structural System, and its Probable Erectional Procedures.* Syracuse, NY: Syracuse University Press, 1968.

Halsted, Bryon David. *Barn Plans and Outbuildings.* New York: Orange Judd Co., 1918.

Hubka, Thomas C. *Big House, Little House, Back House Barn.* Hanover: University Press of New England, 1984.

Lindley, James A. and James H. Whitaker. *Agricultural Buildings and Structures.* St. Joseph, MI: ASAE, 1996.

Neubauer, Loren W. and Harry B. Walker. *Farm Building Design.* Englewood Cliffs, NJ: Prentice-Hall, 1961.

Noble, Allen G. *Wood, Brick and Stone: The North American Settlement Landscape Vol. 1: Houses.* Amherst, MA: The University of Massachusetts Press, 1984.

———. *Wood, Brick, and Stone: The North American Settlement Landscape Vol. 2: Barns and Farm Structures.* Amherst, MA: The University of Massachusetts Press, 1984.

Noble, Allen G. and Hubert G. H. Wilhelm, eds. *Barns of the Midwest.* Athens, OH: Ohio University Press, 1995.

Nobel, Allen G. and Richard K. Cleek. *The Old Barn Book.* New Brunswick, NJ: Rutgers University Press, 1995.

Rawson, Richard. *The Old House Book of Barn Plans.* New York: Sterling Publishing Company, 1990.

Schultz, LeRoy G. *Barns, Stables and Outbuildings: A World Bibliography in English, 1700-1983.* Jefferson, NC: McFarland & Company, 1986.

Kirk, Malcolm. *Silent Spaces: The Last of the Great Aisled Barns.* Boston, MA: Little, Brown and Co.,1994

Visser, Thomas Durant. *Field Guide to New England Barns and Farm Buildings.* Hanover, NH: University Press of New England, 1997.

Wysocky, Ken, ed. *This Old Barn: Country Folks Fondly Recall — In Words and Photos — The Heart of their Homesteads.* Greendale, WI: Reiman Publications, 1996.